AF324031

Valuation of Coastland Resources

Published by Academic Foundation
in association with the

CENTRE FOR DEVELOPMENT ALTERNATIVES
(CFDA), AHMEDABAD, INDIA

Authors:

INDIRA HIRWAY is the director and professor of economics at the Centre For Development Alternatives (CFDA), Ahmedabad, India. She studied for her Masters at the Delhi School of Economics, Delhi and for her PhD at the University of Bombay, Mumbai. She has published books and articles in several areas—poverty and human development, labour and employment, globalisation and development paradigms as well as environment and development and environment accounting.

SUBHRANGSU GOSWAMI is an environmental engineer and environmental planner. He has been actively involved in various research projects in the field of water, environment and development, at CFDA as well as at the School of Planning, CEPT, Ahmedabad. Currently, he is a doctoral scholar of Planning and Public Policy at CEPT University, Ahmedabad.

VALUATION OF COASTLAND RESOURCES

THE CASE OF MANGROVES IN GUJARAT

INDIRA HIRWAY
SUBHRANGSU GOSWAMI

ACADEMIC FOUNDATION

NEW DELHI

First published in 2007
by

ACADEMIC FOUNDATION
4772-73 / 23 Bharat Ram Road, (23 Ansari Road),
Darya Ganj, New Delhi - 110 002 (India).
Phones : 23245001 / 02 / 03 / 04.
Fax : +91-11-23245005.
E-mail : academic@vsnl.com
www. academicfoundation.com

in association with:

Centre For Development Alternatives (CFDA), Ahmedabad.

Cataloging in Publication Data--DK
        Courtesy: D.K. Agencies (P) Ltd. <docinfo@dkagencies.com>

        Hirway, Indira, 1946-
          Valuation of coastal resources : the case of mangroves
        in Gujarat / Indira Hirway, Subhrangsu Goswami.
            p.   cm.
          Includes bibliographical references (p.   )
          Includes index.
          ISBN-13: 9788171885961
          ISBN-10: 8171885969

          1. Mangrove forests--Valuation--India--Gujarat. 2.
        Coastal forests--Valuation--India--Gujarat. 3. Mangrove
        ecology--India--Gujarat.   I. Goswami, Subhrangsu,
        joint author. II. Title.

        DDC 577.698 095 475   22

Typeset by Italics India, New Delhi.
Printed and bound in India.

CONTENTS

List of Figures, Tables and Annexures

FIGURES

TABLES

ANNEXURES

Acronyms

CFDA	Centre for Development Alternatives
EISC	Environmental Information System Centre
FSI	Forest Survey of India
GEC	Gujarat Ecology Commission
GEERF	Gujarat Ecological Education and Research Foundation
GO	Government Organisation
GUIDE	Gujarat Institute of Desert Ecology
ICEF	India Canada Environment Facility
ISRO	Indian Space Research Organisation
LPG	Liquid Petroleum Gas
MNP & S	Marine National Park and Sanctuary
NGO	Non-government Organisation
ONGC	Oil and Natural Gas Commission
PRA	Participatory Rural Appraisal
REMAG	Regeneration of Mangroves in Gujarat
SAC	Space Application Centre
SAVE	Saline Area Vitalisation Enterprise
SFR	State of the Forest Report
TEV	Total Economic Value
WTP	Willingness to Pay

Acknowledgement

The present study on Valuation of Mangroves in Gujarat is sponsored by Gujarat Ecology Commission, Vadodara. We express our sincere thanks to Mr. Arjun Singh, Chairman of Gujarat Ecology Commission, for sponsoring the study and for providing necessary support. We would also like to thank the former Member Secretaries, Ms. Vilasini Ramchandran and Mr. G.L. Bhagat and present Member Secretary, Mr. L. Chuango for their support. We thank India Canada Environment Facility (ICEF), especially Ms. Jaya Chatterjee and Mr. Ujjwal Chaudhry, for taking interest in this area and providing finance to the study.

We also thank the following persons who extended their help and support to us in carrying out the study:

- Dr. Y.D. Singh, Director, GUIDE.

- Dr. G.A Thivakaran, Scientist, GUIDE.

- Mr. Nishchal Joshi and Ms. Nisha Mistry, Research Associate, GUIDE.

- Mr. Rajesh Shah, Director, SAVE.

- Mr. Kalubhai Dangar, Mr. Hemraj Patel and Mr. Jesang Thakor, Coordinator, VIKAS.

- Ms. Debuben Pandya, Director, MAHITI.

- Mr. Cyril Parmar, Managing Director, Daheda Sangh.

- Dr. Oswin D. Stanley, Mangrove Ecologist, GEC.

- Dr. Jai Sule Pawar, Manager Environment, GEC.

- Mr. R.N. Tripathi, Chief Conservator of Forests, Government of Gujarat.

- Mr. S.P. Jani, Deputy Conservator of Forest, Marine National Park, Jamnagar.

- Mr. Bharat Solanki, Range Forest Officer of Marine National Park, Jamnagar.

- Mr. H.S. Singh, Ex-Director, GEER Foundation.

- Dr. R. Sen Gupta of Gujarat Ecological Society, Vadodara.

- Dr. Shailesh Nayak and Dr. Anjali Bahuguna of ISRO, Ahmedabad.

We are also thankful to the experts who gave their comments and suggestions on the earlier draft of the study during the seminar organised by Gujarat Ecology Commission. We are particularly thankful to Dr. M.N. Murthy, Dr. D.S. Ravindran, Dr. M.G. Chandranath, Mr. Vinay Mahajan and others.

We would also like to thank Mr. Jigar Kantharia, Administrative Officer (Projects) for giving excellent administrative support during the study.

Indira Hirway
Subhrangsu Goswami

Ahmedabad

1

Valuation of Mangroves

A Major Ecosystem on the Indian Coast

Introduction

National income statistics form the basis for measuring and monitoring the performance of an economy, as they provide information about economic growth, structure and sources of incomes, distribution of incomes across regions and across socioeconomic groups etc. However, these data do not include environmental resources adequately, with the result that they fail to provide the required inputs for the formulation of sound economic policies, particularly in the context of sustainable development. Natural resource accounts (NRAs), if compiled properly, can estimate environmental gains and losses in monetary terms for the purpose of providing the right signals for macro policy formulation for sustainable development. These accounts answer the critical questions like, (a) what is the real income coming from the nature, (b) how much the economy/nation borrows from the nature and (c) What should be the right policy with respect to natural resources.

Coastal resources constitute an important component of natural resources, particularly in a country like India, which is surrounded by sea from three sides. Coastal regions, where land and water meet, are unique ecoregions which are also attractive for a large number of economic activities like trade and ports, shipbuilding, fisheries, industries, defence, tourism, human settlements etc. Though these economic activities appear to promote economic growth, they cost in terms of loss of coastal and marine resources. Since no or negligible value is attached to these resources, the right signals are not available for taking appropriate decisions about the economic activities. A major challenge in coastal development is how to manage the balance between the economy and ecology in these regions, and generating statistics on the value of coastal and marine resources is essential for the purpose.

Coastal ecosystems, which are an important part of the natural resource base in India, include several ecosystems like salt marshes, mangroves, coral reefs, sea grass etc. Of these, mangroves are particularly important in view of its pivotal role both from the ecological and economic points of view. Mangroves are defined by botanists as "a tree, shrub, palm, ground fern or a grass, exceeding half a metre in height and growing above the mean sea level in the inter-tidal zone of marine coastal environments or estuary margins in the tropical and subtropical coastal regions." Mangroves, the salt tolerant forest ecosystem, provides a wide range of ecological and economic products and services, and also supports a variety of other coastal and main ecosystems, which again provide several ecological and economic benefits.

Till about 1960s, mangroves were largely viewed as 'economically unproductive areas' and were therefore destroyed for reclaiming land for various economic activities. Gradually, however, the economic and ecological advantages of mangroves have become visible and their importance is appreciated. It is now accepted that these rich ecosystems provide a wide range of ecological and economic products and services, and also support a variety of other coastal and marine ecosystems, which again provide several economic and ecological benefits. The full value of mangroves is, however, still not recognised in most economies, as most of these benefits do not enter the market. Consequently, mangroves have been highly undervalued and neglected ecosystems, resulting in their heavy depletion and degradation in most parts of the world. Recently some scholars have compiled their value in monetary terms in some parts of the world (Hamilton *et al.*, 1989; Christensen, 1982; Aodgeson and Dixon, 1988; Lal, 1990; Ruitenbeek, 1992; Bennet and Reynolds, 1993; Gammage, 1994; Spaninks and Beukering, 1997, etc.). Most of these studies are conducted for developed countries and even when it is for developing countries, the nature of benefits and costs estimated are very much different from the same in India, where mangroves play an extremely important role in the life and livelihood of coastal population. There is therefore, a need to compute monetary value of mangroves in India to understand its importance in our economy.

The main objective of this study is to estimate the monetary value of mangroves, keeping in mind its multiple contribution to the economy and ecology, by using different methods of valuation, for a major Indian State, Gujarat, which has the longest seacoast (about one-fourth of the total Indian coast) in the country. This is a methodological study in the sense that it attempts to develop methodology for valuation of coastal resources in India.

The study, to start with, examines the changing status of mangroves in India over the past decades and analyses the factors responsible for their depletion and degradation over the years. The study shows how the status of mangroves has changed in the different coastal states and what have been the reasons for this. The state-wise analysis throws light on the processes, mainly the pressures of economic growth, which are responsible for the long term losses of mangroves in the country.

The study, then, compiles the economic value of mangroves in Gujarat by using the different methods of valuation, like the replacement method, production method and contingency valuation. The study estimates the contribution of mangroves to Gujarat economy that occurs in multiple ways and at multiple levels.

Mangroves: Critical for Ecology and Economy

Economic Uses

Mangroves have many economic and ecological uses. The economic advantages, however, differ from country to country depending on the status of mangroves on the one hand and the socioeconomic lifestyle of the country on the other hand.

As far as economic uses of mangroves in India, and particularly Gujarat, are concerned, mangroves are consumed by households in coastal areas as fuel wood (charcoal can also be manufactured from mangroves), for construction of boat & houses (timber) and as food (mangroves seeds are used as vegetables in some parts of the state). Mangroves are also used as household medicines (for chicken pox, for injury etc.). Mangroves are also a major source of livelihood of people in coastal areas. Firstly, they are an important source of fodder for animal husbandry in coastal villages. This fodder is believed to be as nutritive as cottonseeds and it helps in improving the fat content of milk. Several households in coastal areas survive, particularly in droughts, by cutting mangroves for selling it for fuel and fodder. Secondly, mangroves are the most productive ecosystem in the world that supports a range of creatures for human consumption, such as, crabs, mudskippers and fish in the waterways of mangroves. It has been estimated that small scale fisheries in mangrove waters in the world produce nearly one million tonnes of fisheries, molluscs, crabs and shrimps annually, that is equivalent to about 1.1 per cent of the world fishery catch (Kapetsky, 1985). Mangroves also support breeding of fish, prawn, shrimp, turtles, crabs and many other sea lives. The seas outside of mangroves are always found to be rich in fishery and aquaculture. Thus mangroves provide direct employment to large number of fisherfolk. Thirdly, mangroves also protect agriculture by preventing saline water from entering agricultural lands, and thereby support livelihood of farmers in coastal areas. Fourthly, mangroves are important as raw material in manufacturing a large variety of products like alcohol and vinegar, gum, honey, and other medicines. Mangroves also support growth of corals, which have important industrial uses.

Pioneer investigations have shown that some of the mangrove vegetation can serve a variety of purposes: (1) as a source of tea; (2) as a cholesterol feed for prawn; (3) as potential source of mosquitocides; (4) for antiviral drug formulation—especially against AIDS and jaundice; (5) as a source of UV—absorbing compounds, and (6) as a source of bacterial biofertilisers (Subramonia Thangam, 1990; Premanathan, 1991; Kathiresan and Ramesh, 1991; Murthy, 1999; Ravi Kumar, 1995; Palaniselvam, 1995; Moorthy and Kathiresan, 1995; Kathiresan, 1995). Some of these investigations are likely to prove the efficacy of the mangrove plants in the aspects of therapeutic preventive and clinical medicines as well as in agriculture. In addition, mangroves are attractive tourist places and therefore a source of developing tourism. They are also important for ecotourism and centres for research and education.

Protection against Strong Winds, Cyclones and Tsunami Waves

One major advantage of mangroves that has been highlighted during the recent super disaster of tsunami waves that engulfed several countries and affected the life of millions of people in Asia, is that mangroves can serve as a major shield to protect coastal people and their property against powerful tsunami waves.

The fishing village of Thirunal Thoppu in Tamil Nadu with 172 households was saved from the devastating tsunami waves only because of the dense growth of mangroves on the coast. Similarly, people on Simeuleu Island in Indonesia, located near the epicenter of the earthquake, were protected from the tsunami waves only because they had a protective buffer of thick mangroves surrounding them. Also, communities lying behind mangroves or intact coral reefs in Maldive Islands suffered less damage and loss of life. (National Geographic April, 2005). National Geographic (2005) also observed that more coastal vegetation might have saved many of the victims drowned or crushed by debris in Khao Lak in Thailand.

Ben Brown, Coordinator of MAP (Mangroves Action Plan) in Indonesia, has blamed the clearing of mangroves on the coast of Indonesia and Thailand, for undertaking shrimp cultivation on a massive scale, for the huge tragedy occurred in these countries. He observed that the fault of this "tragedy of commons" lies with the indiscriminate destruction of coastal vegetation for promoting unsustainable development. Jeff Mc Neely, Chief Scientist of IUCN, has also observed that loss of mangroves to shrimp cultivation by those, who hardly know or care for coastal ecosystems, has been the main reason for the massive destruction of life and property by the tsunami waves. Pisit Chamsnah put it very correctly when he said that, "mangroves protected those who protected mangroves" (Chamsnah, 2005). In short, the scientific verdict of the tsunami tragedy was clear: "the damage was greatest where mangroves were removed, dunes flattered and coral reefs killed."

The experience of 'Super Cyclone' in Orissa in 2002 and the disaster in the islands that killed thousands of people in 2004 in Philippines also indicates that the conservation of coastal vegetation, mangroves, would have saved the life and property of thousands of people. In 1960, in Bangladesh a tsunami wave hit the coast in an area where mangroves were intact. There was not a single human loss. These mangroves were subsequently cut down for shrimp farms. In 1991, thousands of people were killed when a tsunami of the same magnitude hit the same region (Sharma, 2005).

Mangroves provide this 'bio shield' to coastal people as they: (1) break the strength of wind and waves; (2) stabilise sediments and reduce shoreline erosion; (3) regulate flooding, and (4) protect people and property.

To put it differently, mangroves are shoreline stabilisers as they stabilise sediment that has been deposited largely by geomorphological processes. That is why they protect the coast from erosion and damage. They also prevent wind bound salinity from spreading on land and thereby protect agriculture

Other Critical Ecological Functions

The biodiversity protected and supported by mangroves include a wide range of creatures, ranging from small insects like bacteria, and fungi, a variety of fish, prawns, shrimps, etc., to a variety of birds. Mangroves function as a prime breeding and nursery for a large number of fauna. Several creatures survive and flourish under the protection of mangroves and live on the organisms supported by dead leaves and fruits of mangroves. Mangroves serve as custodians of their juvenile stock and as natural wealth (Kathiresan, 1995).

Mangroves protect coral reefs. When mangroves and corals coexist, the prop roots of the mangroves retain the land-derived sediments and prevent their entry into the coral habitat. The presence of mangroves thus protects corals from sedimentation, leading to coral's healthy growth. Also, mangroves help in supporting variety of marine life on dead corals. Similarly, mangroves support a variety of flora—sea weeds, small plants and creepers. It has been estimated that mangroves support a large variety of micro-organisms, plants, invertebrates, fish and prawns, amphibians and reptiles, birds, mammals etc. (Singh 1999). In short, mangroves are one of the most productive ecological systems in the world.

Mangroves are also responsible for several other ecological functions like purification of waters, climate regulation, carbon sequestration and release and watershed services etc. Mangroves reduce salinity of water and also protect land from salinity ingress. Carbon sequestration and climate regulation are seen as an extremely important contribution of mangroves. The relationship between climate change (one of today's leading environmental concerns) and the conservation and development of the world's forest has become a major issue today. Tropical forests, including mangroves, have an important role in regulating carbon dioxide in the global atmosphere through the processes of respiration and photosynthesis, whereby plants absorb carbon oxide/dioxide and store it in their biomass. A major ecological function of mangroves is to serve as carbon sink. This ecological function of mangroves is not merely for the coastal ecology of the region, but it is also for the national and global ecology. In the context of the Kyoto Protocol, mangroves can earn millions of dollars of carbon credit for the country.

In spite of these multiple advantages, mangroves are a badly neglected species in the world today. The following discussion on the status of mangroves in the world and in India reveals this explicitly.

Global Status of Mangroves

The word mangrove is being used in the Oxford English Dictionary since 1613, while the Americans, the Spanish and the Portuguese use the words 'Mangle' which is interpreted from the Haytian Arawak language for the trees and shrubs of the genus *Rhizophora*. Later the word was modified as 'Mangrove' and also included other tree genera growing in intertidal zone. At present the terms mangrove, or mangrove forest or mangrove ecosystem are commonly accepted and have become the synonym of tidal forest.

Scientists theorise that the earliest mangrove species originated in the Indo-Malayan region. This theory is supported by the fact that there are more mangrove species present in this region than anywhere else in the world. Due to many species' unique floating propagules and seeds, early mangroves spread westward, by ocean currents, to India, East Africa, and eastward to the Americas, arriving in Central and South America during the upper Cretaceous period and lower Miocene epoch (66 and 23 million years ago). This may explain why the mangroves of the Americas contain fewer and similar colonising species, while those of Asia, India, and East Africa contain a fuller range of mangrove species.

Today mangroves are observed in about 30 countries in tropical subtropical regions covering an area of about 99,300 sq. km. (Singh,

2000).[1] However, according to the recent study by Spadling *et al.* (1997) mangroves ecosystems are estimated to cover 181,000 sq. kms. The global distribution of mangroves is given below.

Table 1.1

Mangroves in the World

Region	Mangrove Area (km²)	Mangroves Area (%)
South and Southeast Asia	75,173	41.5
The Americas	49,096	27.1
West Africa	27,995	15.5
Australia	18,789	10.4
East Africa and the Middle East	10,024	5.5
Total Area	181,077	100.0

Source: Spalding *et al.* (1997).

South Asia and Southeast Asia have the highest coverage of mangroves, with about 41 per cent area of mangroves of the world lying in this region. This region is followed by the Americas, West Africa and Australia. During the past 50 years, however, over 50 per cent of the mangrove cover has been lost, mainly because of the increased pressure of human activities like shrimp farming and agriculture, forestry, salt extraction, urban development, tourist development and infrastructure. Dams on rivers, contamination of seawaters caused by heavy metals, oil spills, pesticides and other products etc., have also been found to be responsible for the decline of mangroves. Shrimp farming is directly linked to the loss of mangroves in tropical world. About 2000 sq. km. mangroves in Thailand, 1200 sq. km. in Equador, 670 sq. km. in Vietnam, 350 sq. km. in India and 90 sq. km. in Bangladesh have been lost to shrimp farming (Anon, 1998). Table 1.2 gives an estimation of losses in mangroves area in 10 countries in South and Southeast Asia. One finds India experiencing the second highest loss (85 per cent) in the area under mangroves, preceded by Thailand (87 per cent).

---

1. Finlayson and Moser (1991) estimated the total mangrove area of the world at about 140,000 sq. km. while UNDP (1995) has estimated it as 240,000 sq. km.

Table 1.2

Estimated Loss of Original Mangrove Area in South and Southeast Asia

Countries	Loss of Original Mangroves Area (%)
Bangladesh	73
Brunei	17
India	85
Indonesia	45
Malaysia	32
Myanmar	58
Pakistan	78
Singapore	76
Thailand	87
Vietnam	62
Unweighted Average	61

Source: Based on country data available in WRI, 1996.

Mangroves in India

India is rich in coastal marine biodiversity along its coastline of 7,500 km., including the exclusive economic zone of 2.02 million sq. km. Including the coast of Lakshadweep and Andaman and Nicobar Islands, the coastline of India measures about 7516.6 km, which is distributed among nine coastal States and four Union Territories. This coastal region, which fall within the bounds of the tropics, supports some of the most productive ecosystems, of which mangroves is an important ecosystem. As per data furnished by the concerned State Governments in 1987, the total area of tidal forest in India was about 6,740 sq. km, constituting about 7 per cent of the world's area under mangroves (Anon, 1987). But the ground scenario was quite different, as some of the mangrove forests had already disappeared due to human activities. The actual coverage of mangroves, as per the Forest Survey of India (FSI),[2] was much less, about 4481 sq. km. in 1987. In 2001, the year for which the latest data from the FSI are available, the mangroves covered about 4,481 sq. km. Of these, about 59 per cent of mangroves are found on the east coast, 23 per cent on the

---

2. Till 2001, Forest Survey of India, Dehradun, made independent assessments of mangrove cover eight times at interval of every two years using remote sensing technology.

west coast and the remaining 18 per cent on the Andaman & Nicobar Islands.

The highest mangrove area is in West Bengal (2,081 sq. km.), followed by Gujarat (911 sq. km.), Andaman-Nicobar (789 sq. km.), Andhra Pradesh (333 sq. km.) and Orissa (219 sq. km.).

Table 1.3

State-wise Mangroves Cover in India

(in sq. km.)

States/Union Territory	Estimated by States**	Estimated by Forest Survey of India (1987-2001)*							
	1987	1987	1989	1991	1993	1995	1997	1999	2001
Andaman & Nicobar	1190	686	973	971	966	966	966	966	789
Andhra Pradesh	200	495	405	399	378	383	383	397	333
Goa	200	0	3	3	3	3	5	5	5
Gujarat	260	427	412	397	419	689	901	1031	911
Karnataka	60	0	0	0	0	2	3	3	2
Maharashtra	330	140	114	113	155	155	124	108	118
Orissa	150	199	192	195	195	195	211	215	219
Tamil Nadu	150	23	47	47	21	21	21	21	23
West Bengal	4200	2076	2109	2119	2119	2119	2123	2125	2081
Total	6740	4046	4255	4244	4256	4533	4737	4871	4481

Source: *- State of Forest Reports prepared by Forest Survey of India (1987–2001).
    **- Anon (1987).

Biodiversity of Mangrove Ecosystem in India

Since the mangrove ecosystem is little understood for its biodiversity in India, the knowledge on occurrence and distribution of mangrove species is inadequate. The Ministry of Environment and Forests (Govt. of India), New Delhi has recently documented the mangrove biodiversity in India, through a funded project that was carried out by Dr. K. Kathiresan of Centre of Advanced Study in Marine Biology (Annamalai University).

Dr. Kathiresan has observed that the east coast and Andaman & Nicobar Islands are richer in biodiversity than the west coast. The mangroves in India comprise 69 species, under 42 genera and 27 families. However, a study has shown that 38 species of mangroves belonging to 21 genera and 18 families are found in Andaman and Nicobar alone, (Mall *et al.*, 1985). Naskar and Mandal (1999) have reported 35 true mangrove species from Indian sub-continent. Out of this, 13 species of mangroves are found in the west-coast of India (Singh *et al.*, 1985).

Many mangrove species are either endangered or rare or endemic. A recent workshop held in Goa has summarised 60 species of plants existing in Indian mangrove habitats. Of these, 12 plant species are critically endangered, 42 species are endangered, and 3 species are vulnerable which deserve immediate attention. Like wise, there are 23 species of algae associated with mangroves; of which, 2 are critically endangered and 12 are endangered. Thus mangrove ecosystem as a whole is in vulnerable state all over India.

Mangroves support rich faunal resources too (Rao, 1991). Among invertebrates, more than 500 species of insects and Arachnida, 229 species of crustacea, 212 species of molluscs, 50 species of nematodes, and 150 species of planktonic and benthic organisms are known from Indian mangroves (Gopal and Krishnamurthy, 1993). Where as vertebrate fauna are represented by 300 species of fish, 177 species of birds, 36 species of mammals and 22 species of reptiles (Gopal and Krishnamurthy, 1993). The respective totals of faunal species are particularly high in the Sundarbans (1434) and in Andaman & Nicobar Islands (914).

Biogeography of Indian Mangroves

There are three different types of mangroves in India *viz.*, deltaic, backwater-estuarine and insular categories. The deltaic mangroves occur on east coast (Bay of Bengal) where the mighty rivers make the deltas. The backwater-estuarine type of mangroves exist in the west coast (Arabian) which is characterised by typical funnel-shaped estuaries of major rivers (Indus, Narmada and Tapti) or backwaters, creeks, and neritic inlets. The insular mangroves are present in Andaman & Nicobar islands where many tidal estuaries, small rivers, neritic islets, and lagoons which support a rich mangrove flora (Gopal and Krishnamurthy, 1993).

In general, mangroves of India can be divided into two parts, namely, mangroves of west coast and mangroves of east coast. The mangroves have a vast existence on the east coast of India due to the nutrient-rich alluvial soil formed by the rivers—Ganga, Brahmaputra, Mahanadhi, Godavari, Krishna and Cauvery and a perennial supply of freshwater along the deltaic coast. But, the deltas with alluvial deposits are almost absent on the west coast of India, only funnel-shaped estuaries or backwaters are present.

Mangroves of West Coast

The west coast mangroves of India are distributed over five maritime states namely, Gujarat, Maharashtra, Goa, Karnataka, and Kerala.

1. Mangroves of Gujarat Coast: Gujarat is the north western state of India and the total length of the coastline, facing the Arabian Sea is about 1600 km. Gujarat has got the second largest area (911 sq. km.) of mangroves in India (FSI, 2001). Based on the geography and geomorphology, the Gujarat coastal zones may be divided into five regions namely: 1. The Rann of Kachchh. 2. The Gulf of Kachchh. 3. The Gulf of Khambhat. 4. The Saurashtra coast. and, 5. The South Gujarat coast. Out of these five regions, the Gulfs constitute the major mangrove zones of the Gujarat Coast.

Rann of Kachchh is the extreme western part of India and is divided into two regions i.e. Great Rann, which covers about 10,500 sq. km. and Little Rann, which covers about 3,000 sq. km. Both these regions are saline desert along with arid climate; though during monsoon months i.e. July to September, this area is inundated with the upstream seawater flow along with tidal ingress from several minor creeks, water ways and canals. The mangroves on the Rann of Kachchh are poor along the Kori creek (RSAM, 1992). In the Gulf of Kachchh, dense mangroves are observed around the Patre creek, the Dide kabet, Valsura, Navlakhi and Kandla and near Mundra jetty. Patches of sparse mangroves are observed near Okha, Poshitra, Pindhara, Dhani, Narara, Sikka, Jindra, Pirotan and near the Jakhau port (RSAM, 1992). The mangrove species like *Avicennia officinalis* and *Rhizophora mucronata* dominate on the Gulf of Kachchh. Gujarat government has declared 455.92 sq. km. area as Marine Sanctuary from Okha to Jodiya. An area of 162.89 sq. km. area at Pirotan Island is notified as 'Marine National Park'. This 'Marine National Park' is important ecologically, with its unique coral reefs, mangroves and other interesting marine flora and fauna.

Saurashtra coast or Kathiawar consists of tidal flats, minor estuaries, embayments and beaches. Rows of dunes are also present along the Saurashtra coast. Gulf of Khambhat area is characterised by the estuaries, like the Sabarmati, the Mahi, the Kim, the Dhandhar and the Tapti; extensive mudflats, dunes, scattered sandy beaches makes this Gulf very much diverse ecotype. On Saurashtra coast, mangroves occur only in sparse patches along the creeks on the inter-tidal mudflats along the Jafarabad creek and the Buthrani creek. In the Gulf of Khambhat, mangroves are distributed along the coast near the Mahi, the Dhadhar, the Narmada, the Kim and the Sena rivers. A small patch of dense mangroves was found on the Aliabet island. In South Gujarat, mangroves exist near the mouth of the Kolak estuary and a small creek near Umargam (RSAM, 1992).

2. *Mangroves of Maharashtra Coast:* The total length of the Maharashtra coast, characterised by several pocket beaches flanked by rocky cliffs, is estimated about 720 km. long. The Maharashtra coastal zone falls under five districts from South to North namely, Thane District, Sindhudurg District, Ratnagiri District, Raigad District and Bombay District.

During the last 25 years the Maharashtra coast lost about 40 per cent mangrove due to different anthropogenic pressures. The Satellite Imagery data shows that the mangrove area is only 118 sq. km. on the coast, located mainly in the mouths of the rivers like Vashishti, Thane and Vaitarana. The areas of mangroves are Ratnagiri, Raigad, Thane, Bombay, Sindhudurg, Mahim, Elephanta Island, Waghotan, Rajapur, Dharamtar, Vasai, Shastri, Vikroli etc. One also observes degradation of mangroves in the state.

3. *Mangroves of Goa Coast:* Goa is situated in the Central-West Coast of India, facing the Arabian Sea and extended North to South. The total length of the coast line of Goa is approximately 120 km. The inter-tidal zones of seven minor estuaries are mostly flanked on both sides by the rocky cliffs and formed with silty-sand and silty-clay along with abundant organic matters. The minor estuaries in this coastal zone are: The Terekhol, The Chapora, The Mandovi, The Zuari, The Sal, The Talpona and The Galgibag. The major mangrove zones are extended in the Zuari estuary and Mandovi estuary and the other sporadic mangrove patches are distributed in the remaining four estuaries and the Kumbharjua cannal connecting Mandovi and Zuari

estuaries. Chodan (Chorao) Island near the village of Chodan (Chorao) is a protected mangroves area. This mangrove forest has been declared as Dr. Salim Ali Bird Sanctuary. It harbours good mangrove flora, comprising about 12 mangrove species and 8 associated species. There are no major anthropogenic threats in this mangrove forest.

4. *Mangroves of Karnataka Coast:* Karnataka state coastline is about 320 km. long and is overshadowed by the Western Ghats mountain range. There is a narrow strip of varying width between the mountain and the Arabian Sea the average width being about 20 km. The coast is lined with sandy and occasionally rocky shores.

Mangroves in Karnataka are of fringing type, found in the intertidal regions along the estuaries, backwaters, islands and other protected areas. As per the records of the state government, mangroves cover an area of 6,000 hectares, of which 1,000 hectares are in Uttara Kannada district alone. However, the FSI survey (2001) has found the area under mangroves much less, only 200 hectares. In general, the mangroves are only sparsely distributed along the Karnataka coast. The dense mangroves vegetation is present mainly at Coondapur, particularly in the confluence zone of the three rivers namely, Chakra, Kollur and Haladi, just before they open into the Arabian Sea (Paretta, 1993).

5. *Mangroves of Kerala Coast:* The length of the Kerala coast is about 560 km., extending from north to south parallel to the 'Western Ghat'. The higher population density on the Kerala coast has resulted tremendous pressure on the natural ecosystem. Mangrove forests were cleared to reclaim land for urbanisation, construction of harbours & ports, prawn farming, coconut plantation and rice-fish culture. Mangroves in Kerala was stretching on about 1,000 sq. km. area a century ago, but it is now reduced to just about 17 sq. km. in isolated bits at Kumaragom, Dharmadom, Chettuva, Nadakavu, Pappinisseri, Kunjimangalam, Chateri, Veli etc. (Basha, 1991). No evidence of present mangroves cover is available for Kerala in the latest State of Forest Report (FSI, 2001). It may be due to its scattered location and highly degraded quality.

Mangroves of East Coast

The mangrove ecosystem of the East Coast of India is mostly deltaic type and distributed in five major deltas as well as estuarine mouths of four maritime states, *viz.* Tamil Nadu, Andhra Pradesh, Orissa and West Bengal.

1. Mangroves of Tamil Nadu Coast: The coastline of Tamil Nadu extends about 950 km., with about 46 big and small rivers. All these rivers carry freshwater and silt particles from the upper reaches and discharge in the coastal zone. Mangroves in the Tamil Nadu are located in the Cauvery delta complex, in Pichavaram, Muthupet and Chattram areas. The common dominant mangroves are *Rhizophora apiculata, Rhizophora mucronata,* which attains 5-7 m height. Occasionally one also finds other species like *Sonneratia apetala, Avicennia marina, Avicennia officinalis, Bruguiera cylindrica, Ceriops decandra, Aegiceras corniculatum* and *Lumnitzera racemes.*

Pichavaram mangroves are situated at about 250 km. south of the city of Madras, on the south-east coast of India. It is located in the Vellar-Coleroon estuarine complex and has many islands separated by intricate waterways. It covers an area of about 400 hectares and is traverced by a large number of channels and creeks, which connect the Coleroon estuary in the south and Vellar estuary in the north. The dominant mangroves are *Rhizophora apiculata, R. mucronata,* which attain 5-7 m height. The Cauvery river is the main riverine system of the Thanjavar District, with the following cannels *viz.* Paminiyar, Korayar, Kilaithangi, Kanthapridan, Marakakorayar and Valvanar. In this area *Avicennia marina, Excoecaria agallocha, A. corniculatum* are the dominant mangrove flora.

The Gulf of Mannar Marine Biosphere Reserve along the coast of Tamil Nadu, is first of its kind in India and southeast Asia. It is situated in the Indian part of the Gulf between India and Sri Lanka. It covers an area of about 10,500 sq.km., running southeast and parallel to the main coastline to a distance of about 170 nautical miles. It is an area of about 21 islands from the northern most Pamban to Tuticorin. The total island area is about 555 ha. The Gulf of Mannar and the islands possess unique mangrove vegetation along with other flora and fauna. The vegetation consists of species belonging to *Rhizophora, Avicennia, Bruguiera, Ceriops, Lumnitzera* etc.

Although mangroves are reported from majority of the islands, this vegetation of Manalli is striking for its luxuriance and diversity. They are not very tall trees, as the height is perhaps reduced due to strong winds lashing here perennially and with greater velocity during monsoons, periodical cyclones etc. The species include plants of *Avicennia officinalis, Excoecaria agallocha, Bruguiera cylindrica, Ceriops tagal, Lumnitzera racemosa.*

2. *Mangroves of Andhra Pradesh Coast:* Total length of Andhra Pradesh coastline is about 1014 km., which is rocky and sandy in ,any parts. Mangroves are present in the deltas of the rivers Godavari and Krishna. Dense mangrove vegetation is found towards the coast rather than on shore land because of the dense branching network of creeks which exists towards the coast (RSAM, 1992). There is more mangrove vegetation on tidal flats on the western side of the Krishna delta than on its eastern side. Dense mangroves are also seen over recent sand/mud spits on the Nizampatnam bay (RSAM, 1992). Sparse mangroves are found on the eastern side of the Krishna delta. The mangrove species like *Rhizophora spp. Avicennia marina, Avicennia officinalis, Ceriops decandra, Bruguiera gymnorrhiza, Lumnitzera racemosa, Sonneratia apetala, Excoecaria agallocha, Acanthus ilicifolius, Suaeda nudiflora* are reported from this area. From Krishna delta *Rhizophora mucronata, Rhizophora apiculata, Bruguiera gymnorrhiza, Excoecaria agallocha, Lumnitzera racemosa, Suaeda monioca, Suaeda nudiflora, Suaeda maritime, Salicornia brachiata,* and several other species are also reported.

3. *Mangroves of Orissa Coast:* The total length of the coastline in Orissa is about 430 km. and this coastline is mainly in the depositional stage with the sediments carried down by the river Mahanadi and other two minor rivers *viz.* Brahamani and the Baitarani. All these rivers form deltas in the Orissa coast, in Bhitarkanika.

The mangroves near the mouth of the Mahanadi river form a creek network consisting of Luna, Jambu, Kharnasi, Khola and Batighar jora creeks. The creeks are arranged parallel to the coast, inundated by daily tides. The Bhitarkanika mangroves are luxuriant due to the beneficial influence exerted by the Brahmani and the Baitarani rivers and their distributaries and creeks upon the terrain. On Balasore coast, there is no influence of fresh water inflow except in the Dhamra river mouth and hence the salinity level remains high except in rainy months (RSAM, 1992). Species like *Avicennia alba, Avicennia officinalis, Excoecaria agallocha, Heritiera fomes, Sonneratia apetala, Rhizophora mucronata, Rhizophora apiculata, Ceriops decandra, Bruguiera parviflora, Aegiceras corniculatum, Phoenix paludosa* and *Porteresia coarctata* are the dominated species of Bhitarkanika.

4. Mangroves of West Bengal: The Sunderbans in West Bengal has the largest area of 2081 sq. km. (FSI, 2001), which forms the largest block of mangroves of the world taken together with Bangladesh. The area recorded was much more (4,200 sq. km.) in Sundarbans of West Bengal than in Bangladesh in the recent past (Krishnamurthy *et al.,* 1987). Sundarban is the only mangrove forest of the world having among its denizens, the famous Royal Bengal Tiger *(Panthera tigris)*. The name 'Sunderbans' may be meant to take either from the Bengali name 'Sundari' tree *i.e. Heritiera fames* or beautiful *(Sundar)* forests or it may mean both (Gopal and Krishnamurthy, 1993). The area is a unique highly productive mangrove ecosystem and is the richest mangrove repository of India.

The Sundarbans mangrove ecosystem is spread over the estuarine tracts of the Ganga-Brahmaputra system. The total geographical boundary of the Indian part of the Sundarbans is 9630 sq. km., which is spread over the southern parts in the districts 24-Parganas (South) and 24-Parganas (North) of the West Bengal. Six principal estuarine rivers *viz.,* Baratala, Saptamukhi, Thakuran, Matla, Gosaba and Herobhanga are connected with Bay of Bengal in their southern extremities. These tidal estuarine mouths are sometimes 6 km. broad, which carry the tidal seawater and inundate these Sundarbans mangrove forest in regular intervals. Most of the estuarine rivers of Sundarbans have lost their earlier connections with the river Ganga and their tidal water turned saline. As such the virginity of these mangrove forests have been torned. The area falls under the inter-tidal zone and has a tropical humid climate.

Based on the physiographic characters of different delta lobes, the Sundarbans Biosphere Reserve can be divided into six zones namely, Sagar-Mahisani-Gharamara-Sand group of islands at the estuarine mouth of Hoogly river, Mahisani island on west and Thakuran river on east, Zone between the river Thakuran and Matla, Core area of Sundarbans Tiger Reserve, Buffer area of Sundarbans Tiger Reserve, and the area lying east of Matla river. The Sunderbans is famous for its richness and diversity of mangrove vegetation with dominant species *viz.*, *Avicennia spp.*, *Sonneratia spp.*, *Excoecaria agallocha*, *Rhizophora apiculata*, *R. mucronata*, *Bruguiera gymlorrhiza*, *Ceriops decandra*, *Phoenix paludosa etc.* (RSAM, 1992). It is the home of number of endangered and threatened species of plants. All mangrove flowers are nectar bearing. Sundarbans produce 50-60 m tonnes of honey every year. Twenty-nine (29) species of mammals, 144 species of birds, 55 species of reptiles and 7 species of amphibians have been recorded in Sundarbans. The endangered animal species are Royal Bengal Tiger, Fishing Cat, Gangetic Dolphin and King Crab.

Nearly 24 lakhs people are dependent on the area under Sundarbans Biosphere Reserve. The reclaimed inter-tidal lands are single croplands for paddy, where winter-irrigation is not possible from saline tidal waters. Hence, people have to depend on aquaculture, fishing, honey collection and woodcutting even bearing the man-eating tigers.

5. *Andaman and Nicobar islands*: These islands harbour a rich diversity of mangroves. Dense mangroves are found on these islands along the creeks, near bays and lagoons with dominant species— *Rhizophora mucronata*, *Bruguiera gymnorrhiza*, *Avicennia spp.*, *Ceriops tagal* etc. Here mangroves occupy an area of about 770,00 ha (RSAM, 1992). Bagla and Menon (1989) gave a figure of around 66,261 ha of mangroves in the Andaman-Nicobar islands. However, according to FSI total area under mangroves in the area is about 789,00 ha as on 2001.

Depletion and Degradation of Mangroves in India

Though mangroves in India does not show any significant decline in the recent years in most states, they have declined considerably during the past 4-5 decades in all the states. This has been due to anthropogenic or societal pressures such as land reclamation for settlements, overexploitation, cattle grazing, and harvesting for medicines, timber, and food, fishing etc. and due to natural factors, which are beyond human control. Aquaculture, reclamation of swamps for paddy cultivation and salt production etc. appear to be the main reasons for degradation of mangroves in many states.

As seen above, India has already lost about 85 per cent of original mangrove area that actually existed a century ago (WRI, 1996). The

National, Remote Sensing Agency (NRSA) has recorded a decline of 7,000 ha of mangroves in India within six-year period from 1975 to 1981. The fast destruction and degradation of the mangroves have already caused coastal erosion and fall in fishery resources. A coastal place in Tamil Nadu (Vedaranyam), has lost 40 per cent of its mangrove area with a reduction of 18 per cent of fishery resources within 13 year-period from 1976 to 1989 (Padmavathi, 1991). If the trend is not reversed, mangroves will get completely wiped out from our country (Kathiresan, 1995).

The mangroves in India experience several threats in the different maritime states. The most significant threat is the growing human pressure on mangrove-resources. However, frequently occurring natural calamities like cyclone, storms, surges and floods are also posing a threat to mangroves. (Kathiresan, 1995).

Different researchers at different point of time have studied damages to the mangroves forest in India. Most of them argued that mangroves in India are endangered by hostile habitat and human abuse. For example, in the western part of the Sunderbans, large areas are settled and cultivated by people and very little natural mangrove forest remains. Further more, reduction in the inflow of freshwater as a result of the construction of the Farakka Barrage upstream, resulted into reduced regeneration of mangrove seedlings (Scott, 1989). In the Andamans and Nicobar about 10,000 ha have reportedly been cut since 1960, mainly for fuel (Bagla and Menon, 1989). In Orissa, large area of mangroves are cleared in the Hatamundia reserved forest for aquacultural purposes. Mangroves have also been degraded and cleared near Karanjmal and near Pradip port on the mouth of the Mahanadj river (RSAM, 1992). In Andhra Pradesh, the mangroves are destroyed for prawn culture, salt manufacturing, domestic uses, and also by cyclonic effects and due to replacement of *Casuarina* plantations by farmers (Jayasundaramma *et al.,* 1987). Cattle grazing is a major problem for destruction of mangroves especially in Pichavaram of Tamil Nadu as well as in some parts of Gujarat. The factors responsible for degradation of mangroves in Arabian sea are basically agriculture, constructions, urban development and industrialisation (Untawale *et al.,* 1991)

In Gujarat State, there is an excessive grazing by camels in the Great Rann (Scott, 1989); while in the Gulf of Kachchh, loss of mangroves is as a consequence of cutting for fuel and timber. In the past, mangroves trees were taller and denser. Many areas are now reduced to shrubby growth due to grazing (Chavan, 1985; Scott, 1989). Mangrove destruction proceeds almost unhindered in Karnataka, Goa and Maharashtra. Mangroves on the Karnataka coast are being seriously damaged by pollution and felling. Mangroves in Goa have been polluted by oil. The oil has affected mangroves near Elephant Island, Bombay also. The mangrove lands are privately owned in almost all parts of Maharashtra. Many people have raised the level of mangrove land after clearing the vegetation to stop the inflow of high tide water for farming or construction of buildings (Bhosale and Mulik, 1991). In some areas of Maharashtra, roads are being constructed right in the mangrove areas (Bhosale and Mulik, 1991).

The large coastal lagoons of Kerala have suffered from uncontrolled urban development and mangroves have been felled in this region. Also, large areas of mangroves cleared for agriculture and the construction of extensive bunds have drastically altered the ecology of the swamps (Scott, 1989). As a result, mangroves are now reduced to a few discrete stands, confined to some small pockets of the Kerala backwaters.

Conservation Measures

Conservation measures have been initiated in most states since the late 1980s and protection has been provided to mangroves. Realising the importance of mangroves, the Government of India initiated efforts for their conservation and management in the 1990s. Following the Coastal Regulation Zone[3] (CRZ) Notification in 1991 that prohibited development activities and disposal of wastes in coastal areas, the Government declared mangroves areas as ecologically sensitive areas under the Environment (Protection) Act, 1986 and banned their exploitation. The Ministry initiated a Plan-scheme on conservation and management of mangroves and coral reefs in 1986

---

3. Coastal Regulation Zone (CRZ) Act, 1991 was enacted by the Government of India to protect Indian coast from degradation. The area influenced by tidal action up to 500m from High Tide Line (HTL) and the land between the Low Tide Line (LTL) and the HTL has been declared as Coastal Regulation Zone (CRZ). As per classification system of the CRZ, coastal zone can be divided into CRZ-I, II, III and IV categories. The CRZ-I zone includes ecologically sensitive areas, mangroves, coral reefs area close to breeding ground of fish and other marine life, areas of outstanding natural beauty and Marine Protected Areas. This zone is qualified for strict protection (Singh, 2002).

and constituted a National Committee to advise the Government on relevant policies and programmes.

In 1987, the National Mangrove Committee identified 15 areas (Table 1.4) to start with, for conservation and preparation of management action plan and also identified the nodal academic/ research institutions for the purpose. State Level Steering Committees have been expected to draw action plans for the identified areas. The action plans broadly cover natural regeneration in selected areas, forestation and protective measures. These plans have been implemented with the financial assistance of the Ministry of Environment and Forests. Simultaneously, many research projects concerning with mangroves' have been carried out in academic institutions, with the financial assistance of Ministry of Environment and Forests and various State and Central governmental agencies.

Table 1.4

Areas Selected for the Management of the Mangroves by the National Mangrove Committee

Sr. No.	Location	Area (in Ha.)
1	Northern Andaman	34840
2	Nicobar	
3	Sunderbans, West Bengal	258477
4	Bhitarkanika, Orissa	6000
5	Coringa, Andhra Pradesh	22450
6	Mahanadi Delta, Orissa	5000
7	Pitchavaram, Tamil Nadu	1358
8	Goa	2000
9	Godavari Delta, Andhra Pradesh	6000
10	Gulf of Kachchh, Gujarat	–
11	Coondapur, Karnataka	–
12	Achra/Ratnagiri, Maharashtra	20000
13	Vembanad, Kerala	17924
14	Point Calimere, Tamil Nadu	120
15	Krishna Estuary, Andhra Pradesh	2651

Source: Govt. of India (1997). *Mangroves of India, 1997.*

Under a public litigation in the Supreme Court of India, the court has directed all the States to prepare and implement the CRZ plan. Most of the States have prepared their plans to protect coastal zones from degradation. Government of India also initiated action through State governments to create a network of Marine Protected Areas (MPAs) under the Wildlife (Protection) Act, 1972 to provide protection to critical and important marine ecosystems. However, distribution of the MPAs in India is not uniform. Most of the MPAs are on the east coast and Andaman Islands, with poor presentation given to the other regions. West Bengal, Orissa and Andhra Pradesh have major shares while Karnataka, Kerala and Lakshadweep Islands have yet to identify sites for declaring sanctuaries or national parks in the marine environment. In all, 31 MPAs are designed and they cover an area of about 6,271.21 sq. km., (15.6 per cent of the total coastal wetlands of the country), although this figure is very small when Exclusive Economic Zone of continental shelf is taken into consideration. These MPAs cover about 1.33 per cent of total continental shelf area of the country. There is no MPA in the country, which covers open seawater (Singh, 2002). There are three notified Biosphere Reserves too, which primarily or partially cover marine ecosystems. These are Sundarban Biosphere Reserve, Great Nicobar Biosphere Reserve and Gulf of Mannar Biosphere Reserve.

The acts and policies that cover mangroves are Indian Forest Act, 1927, Forest (Conservation) Act, 1980, National Forest Policy, 1988, Wildlife (Protection) Act, 1972, Environmental (Protection) Act, 1986, Coastal Zone Regulation Act, 1992 and Coastal Zone Management Plans of the State Government. The Forest Department, Maritime Boards, Department of Fisheries, Petroleum Industries, Salt Industries, Indian Navy, Coast Guard, Department of Customs, Department of Tourism and other stakeholders work in the coastal areas and use and/or control its resources. Maritime and Port Policy and the Port Act of States also affect coastal ecology. However, there is no meaningful coordination between coastal/marine environment related policies and developmental policies, and both the sets of policies are designed and implemented independently of each other. Promoting coordination between 'developmental' and 'environmental' policies is a major challenge to policy makers.

Present Study

As observed earlier, it is important to quantify these values in physical terms and in monetary terms. The present study attempts to do this. The specific objectives of the study are as follows:

1. To study the changes in the status of mangroves in Gujarat state during the past two decades or so and to estimate the nature and extent of depletion and degradation of mangroves in physical terms.

2. To compile monetary value of changing status of mangroves using alternative methods.

3. To develop a methodology of computing value of a renewable natural resource in the process, and

4. To infer policy/action implications of the study for improving the status of mangroves in the state.

This book has been divided into six chapters including this first chapter. Chapter 2 presents the changing status of mangroves in the state during the past decades and discusses the causes of depletion and degradation of mangroves in the state, while Chapter 3 presents the approach to valuation of mangroves and describes the methodology of the study. Chapters 4, 5 and 6 present the alternative valuations of mangroves while Chapter 7 presents conclusions and inferences.

2

Changing Status of Mangroves in Gujarat

Mangroves in Gujarat

Gujarat coast is of about 1650 km. in length (21 per cent of India's coast) with 165,000 sq. km. of continental shelf (35.3 per cent of India's continental shelf) and 200,000 sq. km. of Exclusive Economic Zone (9.9 per cent of India's EEZ). It has certain peculiar geophysical characteristics, which are different from the characteristics found in other coastal areas in the country. Most of the coast is characterised by low and uncertain rainfall, relatively high salinity ingress, high wind velocity and poor vegetations. Also, there are wide variations in the geophysical conditions across the different parts of the coast ranging from heavy rainfall in the region in the south to arid climate in the north in Kachchh. The coast can be divided into five major regions based on their geophysical characteristics: Rann of Kachchh, the Gulf of Kachchh, the Saurashtra Coast, the Gulf of Khambhat and the South Gujarat Coast. The growth of mangroves is different in these different regions.

The Rann of Kachchh comprises Little Rann and the Great Rann, which were in the past a shallow inland sea. Coastal wetlands of the Rann broadly comprise high tidal flats vegetations, saline grasslands, network of uplands called Bets and areas with *Prosopis Juliflora* on fringes and Bets. Kori creek in the Great Rann and Surajbari creek in the Little Rann feed tidal water, although seawater reaches to only part of the Rann near the creeks.

Creek areas of Surajbari and Kori, connecting both Ranns, have a different type of environment. They bring important ecotonal effect on the ecology of the area and have high conservation values. Marsh vegetation with mangroves is present along the creeks. The Kori creek area is particularly rich in mangroves making this area the best mangroves area in the state. Dense growth of *Avicennia* is observed in this area, with the height of the tree more than 5 m. This area receives adequate tidal water with silt and minerals from the rivers and streams of Pakistan. These are thus Indus Deltaic Mangroves. Though these mangroves suffered from the cyclone of 1999, they are still one of the best mangroves in the state. About 65 per cent of mangroves of the state are located in this region.

Table 2.1

Mangrove Cover in Jamnagar Coast

(in sq. km.)

	Location	Dense Mangrove	Sparse Mangrove	Total Mangrove Cover
1.	Kori and Adjoining Creek	306.2	337.1	643.3
2.	Lakhpat *Taluka*	2.3	2.1	4.4
3.	Naliya/Abdasa	16.5	5.0	21.5
4.	Mandvi	0.2	0.1	0.3
5.	Mundra	9.5	9.6	19.1
6.	Anjar	7.5	11.9	19.4
7.	Bhachau	2.0	174	19.4
	Total	344.2	383.2	727.4

Source: *Mangroves in Gujarat* (2000), GEER Foundation, Gandhinagar.

Gulf of Kachchh has an area of 7,350 sq. km. which is aligned approximately east west and has the length of about 170 km. and width of 75 km. at the mouth. While the northern coast has tidal flats, the southern coast is characterised by dead and live coral reefs, islands and extensive mudflats. The northern coast or the Kachchh coast has good mangroves on the coast of Mundra, Abdasa and Lakhpat *Talukas*. These mangroves extend up to Kori creek. Some of these forests are good forests of *Avicennia*. Mangroves are also observed near Kandala Port (Anjar and Bhacahu *talukas*), Jongi to Surajbari (Bhachau *taluka*) near Mundra Port (Mundra *taluka*) and in Lakhpat *taluka*. Since these mangroves are not well protected from human activities (unlike Kori Creek and Marine National Park of Jamnagar), they show signs of depletion and degradation.

Table 2.2

Mangrove Cover in Kachchh

(in sq. km.)

	Location	Dense Mangrove	Sparse Mangrove	Total Mangrove Cover
1.	Bets	34.0	24.2	58.2
2.	Jodiya	3.7	14.2	17.9
3.	Jamnagar	32.4	11.0	43.4
4.	Lalpur	1.3	3.3	4.6
5.	Khambhaliya	4.3	9.4	13.7
6.	Kalyanpur/Bhalia	2.9	0.7	3.6
	Total	78.6	62.8	141.4

Source: Singh (2000).

The Jamnagar coast of Gulf of Kachchh has been notified as the Marine National Park and Sanctuary, with the result that these mangroves are observed to be in good conditions. Though the mangroves suffered from degradation up to about 1984-85, there has been a good improvement observed after 1985. The coastal area covered by MNP show good growth of mangroves. Some of the islands in the MNP, such as Pirotan, Jindra, Chhad, Dideka Bet, Bhensid, Kalubhar, Bharidar, Chank Tapu, Noru etc. have dense mangroves. In the non-MNP areas, however, one observes sparse and patchy mangroves. There are sparse and scrubby mangroves on both the sides of Surajbari and small patches on Navlakhi to Kandla coast. There is evidence that suggests that some of these areas had thick mangrove forest in the past (Singh, 1999).

Diversity of mangroves is observed to be very good in this part. Four species/sub-species of *Avicennia* (*A. officinalis, A. Marina, A. Alba* etc.), species of *Rhizophora, Cortops, Aegiceeras* are observed in this region. A survey by GEER Foundation has shown that 20 out of 42 Bets in MNP support mangroves (Singh, 1999).

Saurashtra Coast: The shoreline of Saurashtra and coast is less indented. Sandy beaches form continuous linear strip from Dwarka to Diu. Sand is coarse and pebbly in some parts with rocky coast, while it is fine grained in other parts. The Saurashtra coast is different from the coast of the two gulfs as mudflats here are restricted in many parts. Saltpans, industries and other economic activities predominate on this coast.

Table 2.3

*Mangrove Cover in 1998 on Important Islands of the
Marine National Park*

Sr. No.	Name of the Islands	Area (in ha)
1	Bhaidar	477
2	Noru	495
3	Dhani Bet	575
4	Kalubhar	957
5	Bhens Bet	215
6	Chhad	936
7	Pirotan	39
8	Jindra	418
9	Dede-ka-Mundeka and Chakhadi	973
	Total	5,085

Note: Other islands in the Gulf of Kachchh have mangroves cover less than 100 ha.
Source: Singh (2000).

Mangroves in Saurashtra coast from Dwarka to Khambhat are confined to limited mudflats and creeks near Dwarka and Porbandar, Meyani, Mahuva, Diu, Jafarabad, Bhtharai, Ghogha Jetty, Navabandar, Rohia Bet, Piram Bet, Pipavav Bandar, Navera Bet etc. These mangroves are sparse and scrubby and consist of *Avicennia*. The available records show that in the past there were some good mangroves on this coast, but camel grazing as well as overgrazing by other animals, over cutting, development of industries and other economic activities have led to their decline and degradation. It has been observed that mangroves near Porbandar (Harshad Mata) and Dwarka declined mainly because of overgrazing. However, some efforts of regenerating mangroves have yielded positive results in some small patches (Singh, 1999).

Gulf of Khambhat: The Gulf of Khambhat is characterised by number of large and small estuaries. The Narmada estuary is live and classified as a salt wedge estuary where fresh water flows predominate. The other rivers, however, do not have good fresh water flows due to the small and big dams construed on the rivers. Gulf of Khambhat is 70 km. wide and 131 km. long located between Saurashtra peninsula and the mainland Gujarat. Extensive mudflats of 6-8 km. are distributed all along the coast of the gulf except along the Narmada estuary. *Avicennia* with stunted growth is sparsely distributed along the coast near the Mahi, the Dhadhar, the Narmada, the Kim and the Sena rivers. A small patch of mangroves is also observed on the Alia Bet. The dominant areas under mangroves are seen near Bhavnagar, Devla (Bharuch), Mangrol, Pardi, Jankhsi and Dandi in Surat.

Historically there were good mangrove forests in this region, and even today there is a good scope for regenerating mangrove in the inter-tidal zone of the Gulf of Khambhat (Sen Gupta *et al.*, 1999).

South Gujarat: The South Gujarat coast is comparatively uniform and is broken by few indentations. Narrow sandy beach is found between the Mindhola and the Purna rivers and it extends up to Daman. To the north of the Mondhola river, the coast is characterised by marshy land. Along the estuaries of the rivers Mindhola, Purna, Ambika, Auranga and Damanganga mudflats and marsh vegetation are present. Numerous small tidal creeks also occur along the coast.

There are good possibilities of growth of mangroves on this coast. There are patches of mangroves observed in several parts, like the mouths of river Kolak, Purna to Damanganga estuaries, Auranga estuary etc. The growth is however, stunted due to overgrazing as well as industries and their pollution, construction of infrastructure etc.

Some available records show that there were good mangroves in the past near Hazira at the north of Tapi river as well as near Dandi, Mangrol, Pardi, Jankhri, etc. in Surat district and a number of places in Navsari and Valsad districts (near Umargam). Though the area under mangroves today is only about 4,600 ha, there is a good scope for regeneration of mangroves in this region.

In short, Gujarat state, which has about 21 per cent of the Indian sea coast, has mangroves spread over an area of 911 sq. km., which comes to about 20 per cent of the national mangrove area. About 90 per cent of mangroves in Gujarat are located around the Gulf of Kachchh while the rest of the mangroves are found in the Gulf of Khambhat and on the South Gujarat coast. The major locations of mangroves in the state are (1) The Ranns of Kachchh, (2) Kachchh coast of the Gulf of Kachchh, (3) Jamnagar coast of Gulf of Kachchh, (4) Saurashtra coast—from Dwarka to Khambhat, (5) Gulf of Khambhat and (6) South Gujarat. As presented in Table 2.4, Kachchh and Jamnagar together have about 848 sq. km. of mangroves, which comes to more than 90 per cent of the total area under mangroves in the state.

Table 2.4

District-wise Status of Mangroves

District	Area Under Mangroves in sq. km.		
	Dense Mangroves	*Sparse Mangroves*	*Total Mangrove Cover*
Kachchh	118	588	706
Rajkot	0	1	1
Jamnagar	28	114	142
Bhavnagar	10	6	16
Ahmedabad	1	1	2
Bharuch	17	11	28
Surat	8	5	13
Navsari	1	1	2
Porbandar	1	0	1
Total	184	727	911

Source: State of Forest Report 2001, Forest Survey of India, Dehradun.

As the Table 2.4 shows, only about 20 per cent of mangroves in the state are dense and the rest are sparse or degraded. Even in Kachchh and Jamnagar districts the percentage of dense mangroves ranges between 20 per cent to 24.5 per cent.

Classification of Mangroves

Mangroves can be classified into five categories based on their location settings and distribution: (1) Onshore mangroves–mangroves growing on the shore, (2) Estuarine mangroves–mangroves found on estuaries of rivers, (3) Deltaic Mangroves–mangroves growing on deltas of rivers, (4) mangroves of the gulf and (5) mangroves on offshore islands. The Gujarat Mangroves can be classified as follows under this categorisation (Singh, 2002):

Onshore Mangroves

Small patches of mangroves along the creek near Porbandar, Miyani, Mahua, Diu, Jafrabad, Buthrani, Ghogha Jetty, Pipavav Bandar in Saurashtra.

Estuarine Mangroves

Stunted and sparse mangroves from near Mahi, Dhandhar, Kim and Sna estuaries; small patch on Alia island at the mouth of Narmada estuary; Tapti, Umargam and Kalak estuaries.

Deltaic Mangroves

Indus deltaic mangroves in and around Kori creek near Pakistan (Pir Sonai, Sugar, Sir, Kharo, Ramaria, Kalichod, Sindhodic and Sethwara islands and creeks near Jakhau.

Mangroves of the Gulf

Gulf of Kachchh (Kachchh, Rajkot and Jamnagar districts): Mundra and Kandla area in Kachchh, Navalakhi in Rajkot, 20 islands (Pirotan, Zindra, Chhad, Dedeka Mundeka, Bhains bid, Bhaider, Noru, Chank, Narara, Dhani, Khara Chusna and Kalubhar) and coast mangroves (Jodiya, Khijadia near Rozi Bandar, Nava Bandar, Bedi Bandar, Mashuri Creek, Singach, Sikka, Narara, Poshitara, Asota) in Jamnagar; Shrubby mangroves near Bhavnagar and Alia bet in the Gulf of Khambhat.

The last category, mangroves on offshore islands are not there in Gujarat. Of these four categories, the quality of deltaic mangroves is supposed to be the best in terms of height and thickness of trees. Gujarat has maximum area under deltaic mangroves (Kori creek area) followed by mangroves under the gulf. (Kachchh and Jamnagar districts as well as some regions in Ahmedabad, Bhavnagar, Kheda and Bharuch districts).

As per the FSI assessment, there has been an improvement in the area under mangroves, from 397 sq. km. in 1991 to 908 sq. km. in 2001. The Table 2.5 also shows that though there has been an improvement in the area between 1991 and 1999, there is a decline in the area during 1999 and 2001. Different experts have different views regarding this decline. While some experts argue that the decline has been statistical, brought about by the changing scale of the remote sensing imageries, the others argue that the decline is genuine caused by the natural and human factors working on mangroves.

Table 2.5

Changing Status of District-wise Mangrove Cover in Gujarat State

(in sq. km.)

| District | FSI Assessment[@] | | | | | | GEER-F |
	1991	1993	1995	1997	1999	2001	1998[$]
Ahmedabad	-	-	-	-	-	2	0.8
Bharuch	7	35	12	13	6	28	14.7
Bhavnagar	16	16	19	20	25	16	15.2
Jamnagar	118	118	118	118	140	142	141.5
Junagadh	-	-	-	-	1	1*	1
Kheda	-	-	-	-	-	-	-
Kachchh	239	242	536	836	854	706	727.4
Rajkot	-	-	-	-	-	1	5.6
Surat	14	8	4	4	4	13	21.9
Valsad	3	-	-	-	1	2**	10.3
Gujarat	397	419	689	991	1031	911	938.4

Notes: Interpretation Scale 1: 1,25,000.
           Interpretation Scale 1: 50,000.

Sources: [@] State of Forest Reports by Forest Survey of India (FSI).
         [$] Bahuguna, Anjali *et al.* 1997.
       [$$] Bahuguna, Anjali *et al.* 1998.
        [#] Singh, H.S. (1999).
        [*] As per FSI report (2001) its in Porbandar.
     [**] As per FSI report (2001) its in Navsari.

There are alternative estimates available of the quality of mangrove forests in the state for different years (Table 2.6). These estimates are not strictly comparable with each other. The estimates of FSI for the year 2001 (FSI has given these estimates for the first time) indicates that of the total 908 sq. km. of mangrove only 182 sq. km. (20 per cent) are dense and the remaining 726 sq. km. (80 per cent) are sparse or open. The share of dense forest is very low in both Kachchh and Jamnagar districts where the area under mangroves is the highest. The SAC estimates of 1992 put the percentage of dense mangroves to 15.20 per cent while the GEER Foundation (1998) put this to 48.4 per cent. The GEER estimates of 1998 and FSI estimates of 2001 are on the same scale and therefore, are comparable with each other. According to these estimates there has been a severe degradation of mangroves during this period.

Table 2.6

District-wise Dense and Sparse Mangrove Cover in Gujarat

(in sq. km.)

District	Dense			Sparse			Total		
	SAC 1992*	GEERF 1998**	FSI 2001***	SAC 1992*	GEERF 1998**	FSI 2001***	SAC 1992*	GEERF 1998**	FSI 2001***
Ahmedabad	-	0.09	1	28.1	0.7	1	14.6	0.79	2
Bharuch	-	5.69	17	27.5	8.99	11	27.5	14.7	28
Bhavnagar	-	9.19	10	32.3	5.96	6	32.3	15.2	16
Jamnagar	24.8	78.65	28	188.3	62.8	114	213	141	142
Junagadh[#]	-	0.65	1	0.6	10.4	0	0.6	1	1
Kachchh	134	344.19	118	575.6	383	588	710	727	706
Kheda	-	-	-	17.5	0	-	17.5	0	-
Rajkot	0	0.98	0	0	4.59	1	0	5.57	1
Surat	-	12.2	8	8.1	9.64	5	8.1	21.8	13
Valsad[##]	-	3.78	1	-	6.5	1	-	10.3	2
Total	158.8	455.42	182	878	483	726	1037	938	911

Note: Data of 1992 was interpreted in a scale of 1: 1,25,000 and data of 1998 and 2001 were interpreted in a same scale of 1: 50,000.

Source: *– Singh H.S., (1994), **– Singh H.S., (2000), ***– FSI, State of Forest Report 2001, #– As per FSI report its in Porbandar, ##– As per FSI report its in Navsari.

Changing Status of Mangroves in the State

For the purpose of valuation it is important to have data on the changing status of mangroves in the state. However, one major problem in examining the changing status of mangroves in Gujarat is the paucity of data. To start with, there are no comprehensive data available on the status of mangroves before 1991 as the Forest Survey of India started collecting the data on mangroves forest only in 1991.

The available data before 1991 are scattered, irregular and do not have any comprehensive estimates. After 1991 there are biannual reports of the Forest Survey of India that provide data on the area covered under mangroves. However, it is only in 2001 that the FSI provided data on quality of mangroves (dense and sparse mangroves) for the first time.

There are two more sources of data on mangroves in Gujarat, namely, the data generated by SAC of ISRO and the GEER Foundation both of which conducted special studies on mangroves. In addition, the forest department also provides data on afforestation/plantation of mangroves on the Gujarat coast. However, the data produced by GEER Foundation, SAC-ISRO and FSI are not strictly comparable with each other, with the result that it is not easy to come to firm conclusions about the changing status of mangroves in the state. The data from the forest department provide information about plantation of mangroves only.

Mangroves between 1960s and 1980s

Though there are no comprehensive time series data on the coverage of mangrove forests in the state for the period before 1991, there is evidence to show that the state has experienced severe depletion and degradation of mangroves in the past.

The first evidence of rich mangroves forest in the state comes from the vegetation map of Khathiawad prepared by the French Institute of Pondicherry in 1960. It shows that there were dense mangrove forest at Kandla, at southern coast of Gulf of Kachchh and western coast of Gulf of Khambhat. Earlier, the *Imperial Gazetteer of India* Vol. XV111 (1908) also had recorded that Navanagar state (Now Jamnagar) had mangrove swamps lined on the shores of the Gulf; and these provided large supplies of fuelwood and fodder to people. There are several regional studies, which have recorded depletion and degradation of mangroves in different regions of the state.

- Ashwini Kumar's study (1996) in Khambhat (Bhal area) shows that there has been a drastic reduction in the area under mangroves from 438 sq. km. in 1870 to 325 sq. km. in 1960s to 100 sq. km. in 1980 and to a mere 13 sq. km. in 1993. According to the author, this decline has taken place due to human activities as well as natural processes.

Table 2.7

Changes in the Area under Mangroves in Khambhat
(Bhal Area) between 1870 to 1993

Year	Area (sq. km.)
1870	438
1960	325
1970	160
1980	100
1987	105
1987-1990	56
1993	13

Source: Kumar, 1996.

- Ashwini Kumar's (1996) study also shows that in non-Bhal areas of the Gulf of Khambhat mangroves have shown a sharp decline between 1975-1983. Also, mangroves, which were present 30 years ago in the villages of Jambusar, are reduced to small patches, with only Nada having some mangroves in the 1990s.

Table 2.8

Status of Mangroves in Jamnagar and Rajkot

Forest/ Division	Taluka	Area (sq. km.)
Jamnagar (including area of Maliyamiyan of Rajkot)	Jamnagar	141.90
	Jodiya	105.00
	Kalyanpur	20.50
	Khambhalia	242.90
	Lalpur	20.00
	Maliamiyan	77.70
Rajkot	Mundra	53.30
	Abdasa	529.50
Area Under Mangroves		1190.80

Source: Working Plan Circle, Vadodara, 1977 (Conservator of forests).

- Studies by Blasco (1977) and Untawale (1985) show that mangroves in Gujarat have depleted at a very fast rate between 1960s and 1980s due to over exploitation by people, natural causes like cyclones, etc.

- The whole coastline from Okha to Navalakhi and Surajbari (i.e. Southern coast of Gulf of Kachchh) in the Jamnagar and Rajkot districts was covered with thick mangroves forest. Now one finds only isolated patches of sparse mangroves in this area. The exceptions are the various Bets (islands), which form Marine National Park and Sanctuary (GEC, 2001).

- The NRSA recorded a decline of 7,000 ha of mangroves in India between 1975 and 1981. Gujarat also has experienced a big decline (Environmental Information System Centre (EISC), 1997).

- A study by Scott (1989) in Kachchh has shown that excessive grazing by camels on the coast has unprotected mangroves in the Gulf of Kachchh. Mangroves were also cut indiscriminately for fuelwood and timber. It has been estimated that 95 per cent of the mature mangrove trees were cut on the coast between 1975 and 1985 (Chawan, 1985 and Scott, 1989).

- In the Gulf of Khambhat there were extensive tracts of mangroves in the 1970s. But these have been reduced to open scrub forest by excessive exploitation (Scott, 1989 and IUCN, 1999).

- A study by Nayak and Bahuguna (2001) has shown that, (1) mangroves in Gulf of Khambhat has experienced severe degradation between 1975 and 1985; (2) the coast between Kandla and Navalakhi had good mangroves in the past (1970s and 1980s), but these are now degraded badly; (3) mangroves in south Gujarat, in Umargam, Vansiborsi and surrounding areas declined mainly due to industrial pollution, and (4) Alia Bet in South Gujarat had rich and dense mangroves about 25 years ago but today they are badly depleted and degraded.

- Jani (2003) has shown that the area covered under MNP in 1984 underwent severe depletion beween 1975–1992. The dense mangrove forests declined from 158.4 sq. km. in

1975 to 21.8 sq. km. in 1982 and to 23.4 sq. km. in 1985, while the sparse mangrove forest declined from 80.1 sq. km. in 1975 to 28.2 sq. km. in 1982 to 10.00 sq. km. in 1985. That is, the total area under mangroves declined from 238.5 sq. km. in 1975 to 33.4 sq. km. in 1985.

Table 2.9

Trend of Change in Mangrove Habitats at the Core Area of the Marine National Park, Gulf of Kachchh

Categories	Area in sq. km.				
	1975	*1982*	*1985*	*1988*	*1992-93*
Mudflats	7.7	125.6	163.2	81.1	79.6
Mangroves	138.5	50	33.4	55.7	61.3
Dense Mangroves	58.4	21.8	23.4	28.6	48.6
Sparse Mangroves	80.1	28.2	10	27.1	12.7
Saltpans	8.4	13.7	17.5	18.4	25.7

Source: Bahuguna *et al.*, 1997. (In proceedings of the workshop on Integrated Coastal Zone Management, 1997).

In short, there is enough evidence to show that there has been severe degradation of mangroves in almost all the parts of the coast in the state. The damage has been particularly severe in Gulf of Khambhat and in South Gujarat.

Depletion and Degradation of Mangroves in the 1990s

According to the Forest Survey of India, Gujarat had 427 sq. km. area under mangroves in 1987. This declined to 412 sq. km. in 1989 and to 397 sq. km. in 1991. During 1991-1999, however, this area has continuously increased, reaching 1,039 sq. km in 1999. It declined thereafter reaching to 911 sq. km. in 2001 (Table 2.5). In short, the area under mangroves declined between 1987 and 1991 but it increased between 1991 and 2001. A careful examination of the data, however, reveal the following:

- The increase is mainly experienced in two districts, Kachchh and Jamnagar, which together have more than 90 per cent of the total mangroves of the state. The area under mangroves increased from 118 sq. km. (1991) to

142 Sq. km. (2001) in Jamnagar and from 239 sq. km. (1991) to 706 sq. km. (2001) in Kachchh. That is, more than 96 per cent of the increase was experienced in these two districts only. Of the rest of the districts, three districts (i.e. Bhavnagar, Surat and Valsad) experienced a decline while Bharuch experienced an increase. Surat experienced severe depletion between 1991–1999, from 14 sq. km. to 4 sq. km.

- It needs to be noted that there has been a significant decline in the area under mangroves between 1999 and 2001, from 1,031 sq. km. to 908 sq. km. implying around 12 per cent decline. However, the two data sets are not strictly comparable as the scales of the imageries are different. It is difficult to say therefore, how much of this decline is statistical and how much of it is real. Experts have argued that this decline is also due to several natural factors as well as human factors (Nayak and Bahuguna, 2001; Bahuguna, 2003 and Y.D. Singh, 2003).[1]

- The highly uneven distribution of mangroves on the coast has continued, as not much progress could be achieved in lagging districts. The potential area identified for the growth of mangroves (Singh, 1999) in Saurashtra, Gulf of Khambhat and South Gujarat could not even be touched! Consequently the share of Kachchh and Jamnagar districts in the total mangrove area increased from 89.92 per cent in 1991 to 93.4 per cent in 2001, implying further concentration of mangroves in these two districts.

- Though the quality data on mangroves are important there are no time series data on the share of dense mangrove forests in the total mangrove area. The FSI has given these data for the first time in 2001. The data show that only 20 per cent mangroves are dense and the rest are sparse. The 1998 estimates of GEER Foundation and the 2001 estimates of FSI are comparable and they indicate a severe degradation of mangroves during the period.

---

1. We held discussion on this with Dr. Bahuguna, Dr. Y.D. Singh and Mr. S.P. Jani of Marine National Park & Sanctuary.

Table 2.10

District-wise Dense and Sparse Mangrove Cover in Gujarat

(in sq. km.)

District	FSI Assessment 2001***			Dense % to Total
	Dense	Sparse	Total	
Ahmedabad	1	1	2	50.00
Bharuch	17	11	28	60.71
Bhavnagar	10	6	16	62.50
Jamnagar	28	114	142	19.72
Junagadh	1[#]	0[#]	1[#]	100.00
Kachchh	118	588	706	16.71
Kheda	-	-	-	-
Rajkot	0	1	1	0.00
Surat	8	5	13	61.54
Valsad	1[##]	1[##]	2[##]	50.00
Total	182	726	911	20.04

Sources: ***– FSI (2001), *State of Forest Report.*
#– As per FSI report in Porbandar.
##– As per FSI report in Navsari.

There seem to be several reasons for this decline. Firstly, mangroves, and particularly those located outside Kori creek and MNP areas are not protected either from camels or from over exploitation by people. Most of these are not under the control of the forest department and are freely available for consumption to all. It is likely that mangroves have been used indiscriminately through out the 1990s particularly during the frequent droughts in the 1990s. Secondly, as the Baseline Surveys conducted by experts in Kachchh and Gulf of Khambhat have shown, several changes have taken place in the topography and location of sands on the seashore in these regions. The changes have created problems for regeneration of mangroves, which has, in all probability, affected their quantity adversely. Thirdly, several economic activities like ports and jetties, infrastructural facilities, industrial parks, expansion of urban settlements have taken place on the coast during the 1990s. These developments, as noted by scholars, have affected the quality of mangroves adversely. And lastly, scientists have observed a severe degradation of mangroves in the Gulf of Kachchh areas the reasons for which are not yet clearly identified. Since this area is relatively protected from camels and human interventions, the reasons are related to natural processes like increasing salinity etc. According to experts this calls for in-depth research by forest ecologists (Nayak and Bahuguna, 2001).

In short, the causes for degradation of mangroves are varied with different factors dominating in different regions at different times. Broadly these causes can be divided into three major groups:

- Overexploitation of mangroves for fodder and fuel-wood by local communities and *maldharis* (i.e. nomads moving with animals from place to place).

- Reduced natural regeneration and

- Diversion of mangrove lands for other uses like saltpans and other industries

Overexploitation of mangroves for fodder and fuel-wood: Over exploitation of mangroves for fodder and fuel-wood by local communities and *Maldharis* is a complex problem with a variety of other sociopolitical causes. With increasing government control over different natural resources and encroachment of common lands and common resources by the rich, local communities have lost their control and responsibilities in the management of common property resources including resources like mangroves. Mangroves have been particularly damaged by the decline of common lands due to encroachment by the rich and frequent droughts in arid and semi-arid areas of the state. This has accelerated the problem by increasing the pressures on mangroves for fodder and fuel-wood. Free grazing by cattle is usually not possible in these areas due to damp muddy conditions. Cattle owners usually cut mangroves and feed them at home. However, camels can easily move in these areas and can even cross the creeks at the time of low tides to eat mangroves. Free grazing by camels can devastate large mangrove areas, as their feet trample the pneumatophores or the breathing roots, blocking respiration and their saliva results in stunted growth of mangroves. Local communities are often helpless in regulating their access to the mangroves, as they have no control over these areas and repeated droughts have aggravated the problem.

Lack of knowledge and awareness of the ecological functions of mangroves in the local communities is also often responsible for the indiscriminate exploitation of mangroves by these communities.

Reduced natural regeneration: Healthy growth of mangroves requires plenty of fresh water flow at the inter-tidal zones. This flow has declined in many coastal areas due to damming of rivers. Reduction in fresh water flows accompanied by increased salinity as well as pollutants in many cases harmed the growth of mangroves in many regions. No major rivers, except Indus with its reduced annual flows, pour fresh water in the Gulf of Kachchh. As a result, only hardy species like *Avicennia marina,* with high salt tolerance, have survived in this region. Similarly, in the Gulf of Khambhat also fresh water inflows from some of the major rivers like Sabarmati and Mahi have declined. Untreated or partially treated effluent discharges from different industries located on the banks of the rivers have aggravated the situation here. The Gulf of Khambhat coast, particularly its eastern shores is mostly industrialised, as it is a part of 'the Golden Corridor' of the state. The Ankleshwar industrial estate is situated at the south bank of Narmada, which is a 'Chemical Industry Zone'. The effluent water is released to Amlakhadi, which finally joins Narmada and then to the Gulf.[2] Similarly the north bank of Narmada harbours fertiliser industry like GNFC near Bharuch, IPCL Gandhar plant, GACL, Birla Copper etc. All these industries release their effluents either to Narmada estuary or to the adjacent coastal waters of the Gulf of Khambhat.

Another factor that seems to have played a role in the degradation of mangroves in the Gulf of Khambhat is the erosion of the shoreline due to violent sea actions. Frequent cyclones have also taken their toll. The cyclones of June 1998 and May 1999 uprooted large number of trees and killed many others by blocking the pneumatophores with thick layers of sand and mud. Free grazing by camels in the mangrove areas causing trampling of seedlings and overexploitation of seeds and

---

2. According to the Gujarat Pollution Control Board , the quality of water at Amalakhadi is highly polluted, crossing all the norms of water, like BOD, COD etc.

twigs by the local communities also lead to the reduced natural regeneration of mangroves.

Diversion of mangrove lands for other uses like saltpans and industries: Diversion of mangrove lands for uses like saltpan and other industries, for ports and jetties, for quarries etc. have damaged mangroves in some areas. In several places, mangroves have been uprooted to reclaim land, dredging has taken place on a large scale, and extraction of minerals have started from quarries—all of which have damaged mangroves. Location of ports and industries in the vicinity of mangrove areas causes damage to the mangroves due to dredging activities as well as construction of roads and other infrastructure. Coastal areas of Kachchh and Janmnagar districts have particularly experienced this diversion of land. Insufficient understanding of the ecological and economic functions of mangroves (mainly due to the lack of its monetary valuation) and political pressures are largely responsible for such diversions.

Restoration and Regeneration of Mangroves in Gujarat

Realizing the importance of mangrove forests, the state government undertook plantation of mangroves right from the 1960s. About 13.60 sq. km. area was covered under mangrove plantation up to the Third Five Year Plan (i.e. 1966). Sinha and Joshi Working Plan (1973–1983) of Kachchh district has documented this plantation on suitable sites on the muddy creeks along the sea coast (Singh, 1999). In 1982, Marine National Park and Sanctuary were formed on the Jamnagar coast with a view of: (1) promoting biodiversity and protecting rare habitats and threatened species; (2) increasing productivity of fisheries; (3) contributing towards increased knowledge of marine sciences; (4) maintaining natural coastal processes, and (5) protecting attractive habitats and species on which sustainable tourism can be developed (Singh, 2002). The marine sanctuary was declared in 1980, which was expanded and converted into Marine National Park and Sanctuary in 1982. The total area covered under the sanctuary is 457.89 sq. km. spread over 42 islands. This area, which was under environmental stress, has been protected against human activities including camel grazing since 1982, with the result that rich mangroves have grown on it, by natural regeneration and plantation. So far mangroves have been planted on about 10,087 ha or 100.87 sq. km. area (between 1983-84 and 1999-00). A recent survey (1998) by GEER Foundation has shown that about 20 of the 42 islands of the MNP have been covered by rich mangroves. However, there is ample scope for further plantation as majority of coastal mudflats in the area are still devoid of vegetation (Jani, 2003).

The department of forest and environment of the state government has undertaken plantation of mangroves in all the major coastal regions, namely, Jamnagar, Kachchh, Junagadh, Amreli, Bhavnagar, Surat and Valsad. So far 149.83 sq. km. area has been covered by the department in the last 17 years which comes to only about 8.8 sq. km. of plantation per year! Maximum plantation is done in the MNP (100.87 sq. km.) and in Jamnagar and Kachchh districts (37.23 sq. km.), adding up to 138.1 sq. km. area, which comes to more than 92 per cent of the total plantation by the forest department in the state. About 1.45 sq. km. areas has been planted in Junagadh, 0.53 sq. km. in Amreli, 1.75 sq. km. in Bhavnagar and about 7.00 sq. km. in South Gujarat.

Recently the state government has designed a three year action plan for mangroves plantation on about 24.00 sq. km. at the cost of 9.97 crores. Most of this plantation will be carried out in the MNP area. The state government has also notified large areas of mudflats as forests for scientific management and increased regulation of anthropogenic activities in the mangrove areas. About 132,440 ha. area (in Jamnagar, Rajkot and Kachchh districts) of coastal mudflats has been notified as reserved forests. However, most of this area is barren and without vegetation. Some of this area has also been transferred to saltpans and cement industry!

In fact, one major problem in regeneration of mangroves has been the allotment of coastal areas to industries, ports/jetties and for other infrastructural development. For example, considerable area from the MNP has been given on lease to salt manufacturing units and considerable area in Kachchh has been allotted to a port (Adani) and to cement industry (Sanghi Cement). Since there is no correspondence between the industrial policy/sectoral policies and environmental policies in the state, no land is easily available for mangroves! Another problem of official efforts has been almost absence of community participation in their programmes. The forest department operates in a top to down fashion, with the result that the

Table 2.11

Mangroves Plantation in Gujarat by Forest Department

(in ha.)

Sr. No.	Year	MNP Jamnagar	Jamnagar Area	Social Forestry, Jamnagar	Kachchh East	Kachchh West	Junagadh Area	Social Forestry, Amreli	Social Forestry, Bhavnagar	Social Forestry, Surat	Rajpipla West	Vyara Area	Total
1	1983-84	7.0	-	-	-	-	-	-	-	-	-	-	7.0
2	1984-85	1.3	-	-	25	75	-	-	-	-	-	-	101.3
3	1985-86	4.1	-	-	25	75	-	-	-	-	-	-	104.1
4	1986-87	17.0	-	-	-	-	-	-	-	-	-	-	17.0
5	1987-88	250.0	-	-	-	150	-	-	-	-	-	-	400.0
6	1988-89	236.6	-	-	150	150	-	-	-	-	-	-	536.6
7	1989-90	102.0	-	-	150	150	-	-	-	-	-	-	402.0
8	1990-91	150.0	-	-	50	75	-	-	5	-	-	-	280.0
9	1991-92	466.0	25	-	100	200	-	-	5	-	-	-	796.0
10	1992-93	600.0	-	-	100	200	-	-	5	-	-	-	905.0
11	1993-94	550.0	-	20	100	250	20	28	20	-	-	75	1063.0
12	1994-95	700.0	74	30	80	170	50	25	10	40	-	25	1204.0
13	1995-96	304.0	65	50	50	150	75	-	15	-	-	-	709.0
14	1996-97	450.0	110	-	50	100	-	-	50	-	70	70	900.0
15	1997-98	1166.0	110	-	50	100	-	-	65	-	70	70	1631.0
16	1998-99	2403.0	110	-	50	100	-	-	-	-	70	70	2803.0
17	1999-00	2680.0	110	-	94	100	-	-	-	-	70	70	3124.0
	Total	10087.0	604	100	1074	2045	145	53	175	40	280	380	14983.0

Source: Srusti (2000), "Mangroves Special Edition", Vol. 28.

local community is neither involved with it nor do they have any stake in the programmes. The results are therefore, likely to be less than satisfactory.

In short, regeneration of mangroves has received a very low priority in the state so far. The forest department has planted mangroves, on a very small scale, and most of it is in the districts of Kachchh and Jamnagar. Mangroves in the rest of the state have been grossly neglected, with about 1 sq. km. of plantation per year in this area by the forest department during the last five years.

REMAG Project

In response to the need for promoting mangroves in the state, Gujarat Ecology Commission, with the help of India-Canada Environment Facility, has initiated a project entitled Restoration of Mangroves in Gujarat (REMAG). The main objective of this project is to promote community based regeneration and sustainable management of mangroves in selected communities along the Gulf of Kachchh and Gulf of Khambhat. The project expects that, (1) there will be increased acceptance and support for community based approaches for natural resource management by communities and by the government and (2) there will be enhanced capacity of participatory organisations to promote and deliver NRM in a pro-poor and pro-women manner.

Seven sites in Gulf of Kachchh and Khambhat have been selected[3] under the project. While selecting the sites, the suitability of site for mangroves regeneration as well as the presence of an active NGO have been kept in mind. The project beneficiaries will be local communities, which will include marginal farmers, agricultural labourers, livestock owners, fishing communities and the general public. The project period is of five years, from 2001–2006 and the total cost is Rs. 1,013.05 lakhs.

The major tasks undertaken in the project in the selected locations include the following:

---

3. The seven sites are (1) Tadatalav in Khambhat *taluka* in Kheda district, (2) Neja and Nada villages in Jambusar *taluka* of Bharuch district, (3) Kantiajal in Hansot *taluka* in Bharuch district, (4) Mahadevpura and Bhangadh villages in Dhandhuka *taluka* in Ahmedabad district and (5) Asirwandh village in Abdasa *taluka* in Kachchh district.

- Orientation-cum-preparatory workshop.

- Conducting PRAs in the selected villages to understand the ground level realities and the local knowledge about mangroves, to create awareness about the importance of mangroves among people; to help them in forming organisation and to motivate them to join the project.

- Baseline survey by experts to assess the feasibility of regeneration of mangroves on the selected sites and to provide technical inputs for regeneration.

- Acquisition of land from the government.

- Nursery making and plantation.

- Gap filling plantation.

- Follow up actions.

It is expected that each of the selected villages will be able to plant mangroves on about 1,000 hectares of area. This success of the first round will accelerate the efforts in the long run.

The REMAG project has thrown very useful light on the various issues associated with regeneration of mangroves. These are discussed below:

Mangroves plantation frequently needs physical works to create favourable conditions for growth of mangroves. It has been noted in several parts of Gujarat that sedimentation or land erosion has changed the topography of the coast in the way that the area is not favourable for growth of mangroves without carrying out suitable physical works. Some times it becomes necessary to dig channels or trenches to allow for water flows for mangroves (Banik, 2002; Wafer, 2002; Deshmukh, 2002 and Singh, 1999). Sometimes there is a danger to the young seedling from the mass of algae being driven in the shores by currents. The algae surrounds young mangrove plants and gets entangled firmly. During low tides when algae dries, it kills the growing tip of the seedling and destroys the plant. It is necessary to remove algae regularly from young plantations, at least during the first two years of growth. Protection of mangroves from camels as well as from overuse by people is also needed urgently, as overuse may destroy upcoming mangroves. It will be important therefore to organise alternative sources of fodder and fuel wood for people. Strict action needs to be taken to prevent camels entering mangroves or people collecting mangroves for own use or for sale.

Community participation through community-based institutions is the best institutional structure for growth of mangroves. Such an institution (grass root organisation of users and growers) will ensure that the mangroves are planted well, protected well and used judiciously. Community based organisations will also ensure involvement of all the stakeholders, particularly the fishing community, households engaged in animal husbandry, agriculture etc. in regeneration.

Some of the problems faced at the ground level by community organisations engaged in the regeneration of mangroves under the REMAG project are worth noting in this context. Firstly, there is always a problem about the availability of land for mangrove plantation. The availability of land, however, primarily depends on the commitment of the government to this activity. It is important that a long-term plan for mangroves is designed in the state keeping in mind its importance to the economy and ecology of the coastal regions. Coastal regions are always attractive for a wide range of economic activities like ports/jetties, trading & shipping, fisheries and fish processing, oil and refineries, salt industry, other industries, navy, tourism etc. However, keeping in mind the value of mangroves, it is important to draw a long term policy for mangroves and allot land for its regeneration. Another major problem is the availability of funds, which also is linked with the state policy. Considering the fact that mangrove plantation generates short term and long term employment on a significant scale and regenerates the coastal ecology, it is important to allocate funds to this activity. The other problems are with respect to seeds (adequate supply), availability of labour (people should be willing to undertake this activity on a required scale), protection from camels, etc. which need to be organised systematically.

The discussion on the status of mangroves in Gujarat can be summed up as follows:

- There has been severe depletion and degradation of mangroves in Gujarat during the past decades.

- The long term depletion and degradation has been relatively more in Saurashtra, Gulf of Khambhat and South Gujarat where: (1) mangroves have been largely unprotected encouraging their indiscriminate overuse; (2) economic activities—ports, industries, urban centres, infrastructural facilities—have encroached upon the coast, and (3) where certain natural processes, such as sedimentation, shifting sands, coast erosion etc. have taken place creating obstacles in growth of mangroves.

- The Kachchh and Jamnagar coast also is not free from degradation. There is a significant decline in mangroves in the area between Kandala and Navalakhi during 1975-1982, from about 700 sq. km. to 133 sq. km. There has been further degradation of these mangroves in the later period. Also, some salt crusts are observed in the mangroves in Kori creek area. These degradations are due to natural causes (may be, increased salinity in the region), which are not yet understood well. This is a serious issue as these large-scale degradations in Kachchh district are not yet well understood by scientists. In fact, this is a big challenge to forest ecologists.

- The efforts of the government for regeneration have been very small as compared to the needs and have been concentrated mainly in Kachchh and Jamnagar districts, creating a wider gap between the status of mangroves in Kachchh-Jamnagar and in the rest of the regions of the state.

- No clear trends are emerging with respect to the changing status of mangroves in the 1990s in the state. There is an increase in the area under mangroves, mainly in the protected areas of Kachchh and Jamnagar, but not much is known about the changing quality of mangroves in the state. Most of the experts, however, agree that there has been deterioration in the quality of mangroves in the state during the last decade.

- The available information on mangroves in the state clearly indicates that there is a decline in the diversity of species of mangroves over the years. This has been largely due to the increasing salinity on the coast, one cause of which is the decline in fresh water flows from rivers.

- Regeneration of mangroves requires scientific inputs (technology), community participation and funds. The ICEF project implemented by GEC is an important project in this context. The results of this project are likely to be useful in designing action in the future.

Potential of Mangroves in the State

What is the potential of mangroves in the state? How many more areas can be covered under mangroves? Is it possible to go back to the situation in 1950s or 1960s?

Mangroves grow only on suitable inter-tidal mudflats, especially at mouths of rivers and along the creeks. Scientists have identified five factors that control growth of mangroves:

- *Climate:* temperature, humidity, rainfall, fresh water flows etc. The quality of mangroves differs in moist regions, sub-humid regions, semi-arid regions and arid regions.

- *Salinity:* Only some varieties grow on saline coasts, and in extremely saline coasts mangroves find it difficult to survive.

- *Tidal Fluctuations*: Mangroves grow in inter-tidal zones. Regular tidal flows are essential for growth of mangroves.

- *Wind Velocity*: Heavy winds are not good for mangroves. Winds help in siltation, (which is favourable to mangroves) but too much wind is harmful to mangroves.

- *Substrata and Soil*: Mangroves grow in depositional environment, in mudflats with low wave energy. Turbulent water and high wind energy prevent regeneration of mangroves, as they destroy shallow root systems of mangroves. That is why deltaic coasts and estuaries are good for mangroves. Fine mudflats composed of silt, clay and high percentage of organic matter favour mangroves.

Keeping in mind these points, H.S. Singh has identified potential areas for mangroves using the remote sensing data provided by SAC,

Ahmedabad (1999). The Table 2.12 presents these data (The details of the potential areas is presented in Appendix A-2.1).

Table 2.12

Potential Area of Inter-tidal Zone Suitable for Mangrove Regeneration

(in ha.)

District	Mangrove Cover	Potential Area in Inter-tidal Mudflat for Mangroves	Percentage Mangroves Cover to Potential Area
Kuchchh	72740	32730	45.00
Rajkot	560	2830	505.36
Jamnagar	14140	11150	78.85
Porbandar	100	40	40.00
Bhavnagar	1520	2000	131.58
Ahmedabad	80	5250	6562.50
Kheda	-	1620	—-
Bharuch	1470	3730	253.74
Surat	2190	2100	95.89
Valsad	1030	2260	219.42
Total	93830	63710	67.90

Source: Singh, H.S. (2000).

The table shows that there is a good potential for growth of mangroves in the state. It is possible to bring about additional area of 637.10 sq. km. under mangroves. This comes to about 68 per cent of the present area! That is, it is possible to raise the area under mangroves to 1,575.40 sq. km. This area comes to about 6.67 per cent of the total coastal wetlands in the state. It is important to note that Kachchh and Jamnagar districts constitute about 67 per cent of the potential. It is also important to note that potential is very high (as per cent to the existing mangroves area) for Ahmedabad, Kheda, Bharuch and Bhavnagar, i.e. the area under the Gulf of Khambhat.

Tapping this potential, however, is not an easy task. It requires a lot of efforts in several directions. To start with, there are areas where natural regeneration of mangroves can be promoted by taking appropriate steps like protection from camels, protection from over cutting, etc. For artificial regeneration the important points are as follows:

- selection of appropriate sites, keeping in mind the factors that influence growth of mangroves;

- selection of right species as it is equally important to select suitable species for plantation, keeping in mind the natural species of the region, salinity of region and other geophysical conditions of the region;

- direct seed sowing (or aerial seed sowing) or nursery depending on the area (direct sowing may work where the sea is quiet and nursery in protected place may be needed when the tidal waves are powerful);

- careful plantation of seedlings, after the seedlings become 5 to 6 months old, in suitable locations and during suitable seasons; and

- gap filling after 6 to 8 months of plantation, when some of the seedlings have died. Regular gap filling is an important component of plantation.

In addition, there are some field level problems, as discussed above, while undertaking actual plantation of mangroves.

To sum up, though mangroves has depleted and degraded in Gujarat over the past decades, there is a good scope for its regeneration. Apart from the area covered by stunted and depleted mangroves at present, there are many other areas where mangroves can grow well. The state government, along with the support of people, can undertake this task of regeneration of mangroves.

Annexure A-2.1

Taluka-wise Status of Mangroves in November 1998

(Area in ha.)

Taluka	Location	Mangrove Dense	Mangrove Sparse	Total of Mangroves Regeneration	Potential Area for Mangroves	Intertidal Zone	Other Classes	Total 5+6+7+8=9
1	2	3	4	5	6	7	8	9
(1) Kachchh District								
Lakhpat	Kori creek and adjoining creeks	30624	33705	64329	19694	43076	0	127099
	Near Sindholi creek	13	50	63	45	108	-	216
	Near Kori creek	-	-	0	76	3614	-	3690
	Near Sugar creek	204	161	365	89	908	-	1362
	Near Kori creek	-	-	0	-	1813	-	1813
	West of Khirsara village	11	3	14	62	195	-	271
		30852	33919	64771	19966	49714	0	134451
Naliya	Near Sindholi creek	20	13	33	84	1182	-	1299
	Near Godia creek	-	-	0	52	22	-	74
	Ganau, Valawadi village	362	131	493	1286	4172	-	5951
	Near Akri moti village	1052	118	1170	2088	4779	-	8037
	Belawanadh, Kosha, Wandh village	114	147	261	386	2658	-	3305
	Jakhau	98	89	187	590	2970	-	3747
	Wandhan & Budia village	-	6	6	-	866	-	872
		1646	504	2150	4486	16649	0	23285

(Contd. ....)

(...contd. ...)

1	2	3	4	5	6	7	8	9
Mandvi	Mandvi, Nawawas, Gundiali	20	13	33	547	504	-	1084
	Siracha	-	-	0	-	253	-	253
		20	13	33	547	757	0	1337
Mundra	Mundra, Navinal	951	943	1894	1684	3792	-	7370
	Luni, Wadala	-	16	16	103	9409	-	9528
		951	959	1910	1787	13201	0	16898
Anjar	Hansari creek	318	499	817	637	4885	-	6339
	Viva village	-	-	0	-	340	-	340
	Wandi	432	686	1118	1438	6845	-	9401
		750	1185	1935	2075	12070	0	16080
Bhachau	Jongi village	-	16	16	61	2260	-	2337
	Amliyara	200	1722	1922	3812	17611	-	23345
		200	1738	1938	3873	19871	-	25682
Total for Kachchh District		**34419**	**38318**	**72737**	**32734**	**112262**	**0**	**217733**
(1) **Rajkot district**								
Maliya	Navalakhi Port	8	139	147	224	107	-	478
	Jajasar, Bagasar	14	100	114	2011	13854	-	15979
	Chach creek	76	220	296	592	1309	-	2197
Total for Rajkot district		**98**	**459**	**557**	**2827**	**15270**	**0**	**18654**

(Contd. ...)

(...contd. ...)

1	2	3	4	5	6	7	8	9
(1)	**Jamnagar district**							
Jodiya	Bets and reefs	3400	2421	5821	1264	4997	5	12087
	Ranjitpur	70	981	1051	4037	4037	-	8651
	Jodiya, Badampur, Kunad	140	262	402	401	401	-	2394
	Balachadi	158	173	331	33	4104	-	4468
		3768	3837	7605	5735	14255	5	27600
Jamnagar	Sachana, Khijadiya,Bedi	1319	537	1856	1106	2816	-	5778
	Vasai Bed, Sarmat	1389	300	1689	173	360	-	2222
	Paileri	499	254	753	674	2981	-	4408
	Between Hadde creek & Pirotan Islands	37	8	45	139	607	-	791
		3244	1099	4343	2092	6764	-	13199
Lalpur	Gangva, Jhankhar, Sikka	134	329	463	162	901	-	4526
Khabhaliya	Chudeswar	367	889	1256	1568	2022	-	4846
	Nana Asota, Beh	67	51	118	506	2239	-	2863
Kalyanpur	Mota Asota, Virpur	42	27	69	580	328	-	977
	Sanosari	-	-	0	-	92	-	92
	Poshitra	243	47	290	503	276	-	1069
		853	1343	2196	3319	5858	-	11373
	Total for Jamnagar district	**7865**	**6279**	**14144**	**11146**	**26877**	**0**	**52172**

(Contd. ...)

(...contd. ...)

1		2	3	4	5	6	7	8	9
(1)	Junagadh district								
	Porbandar	Miyani	-	-	0	40	36	-	76
		Porbandar	65	35	100	-	134	-	234
	Total for Junagadh district		65	35	100	40	170	0	310
(1)	Bhavnagar district								
	Ghogha	Kuda village	6	9	15	-	76	-	91
		Ghogha village	31	3	34	-	-	-	34
			37	12	49	0	76	0	125
	Bhavnagar	Hathab	-	-	0	-	23	-	23
		Bhavnagar, Bhavnagar creek	164	291	455	425	84	-	964
		Akvada	-	22	22	156	-	-	178
		Avania village	16	9	25	510	27	-	562
		Bhavnagar	702	259	961	686	465	-	2112
			882	581	1463	1777	599	0	3839
	Mahuva	Mahuva	-	3	3	178	-	-	181
		Kotada nicha	-	-	0	-	6	-	6
	Talaja	Kotada nicha	-	-	0	44	154	-	198
			0	3	3	222	160	-	385
	Total for Bhavnagar district		919	596	1515	1999	835	0	4349

(Contd. ...)

(...contd. ...)

1	2	3	4	5	6	7	8	9
(1)	**Ahmedabad district**							
Dholka	Vadgam, Vainaj	-	-	0	-	956	-	956
Dhandhuka	Rahatalav village	9	70	79	5253	3407	-	8739
	Total for Ahmedabad district	**9**	**70**	**79**	**5253**	**4363**	**0**	**9695**
(1)	**Kheda district**							
Khambhat	Khambhat, Washa, Rahej	-	-	0	202	11307	-	11509
	Vadgam	-	-	0	198	6017	-	6215
	North of Gulf of Khambhat	-	-	0	1224	7858	-	9082
	Total for Kheda District	**0**	**0**	**0**	**1624**	**25182**	**0**	**26806**
(1)	**Bharuch district**							
Bharuch	Near Ambika river	-	-	0	-	300	-	300
	Alia bet	136	232	368	601	1698	-	2667
		136	232	368	601	1998	0	2967
Jambusar	Near Manekpur village	-	-	0	-	48	-	48
	Near Degam	-	-	0	-	429	-	429
	Near Kavi, Sarod	-	-	0	-	2126	-	2126
	Near Devla	285	261	546	1302	1361	-	3209
	Near Malpura, Sampura, Sigam	-	6	6	385	123	-	514
		285	267	552	1687	4087	0	6326

(Contd. ...)

(...contd. ...)

1	2	3	4	5	6	7	8	9
Hansol	Near Alia bet	-	9	9	193	875	-	1077
Vagra Near Dahej port		-	37	37	354	377	-	768
	Near Asarsa, Islampur, Tankari	-	178	178	390	996	-	1564
	Paniyadra	148	176	324	504	537	-	1365
		148	391	539	1248	1910	0	3697
Total for Bharuch district		**569**	**899**	**1468**	**3729**	**8870**	**0**	**14067**
(1) **Surat district**								
Olpad	Mangrol, Pardi and Jankhri	443	267	710	398	115	-	1223
	Rana, Tena, Sheri Phalia	14	8	22	395	198	-	615
	Danbhari phalia, Dandi	213	387	600	927	1419	-	2946
		670	662	1332	1720	1732	0	4784
Surat	Junagam, Suvali, Tapti estuary	45	11	56	229	1386	-	1671
	Haziara, Sikotarmata	505	291	796	153	1173	-	2122
		550	302	852	382	2559	0	3793
Total for Surat district		**1220**	**964**	**2184**	**2102**	**4291**	**0**	**8577**

(Contd. ...)

(...contd. ...)

1	2	3	4	5	6	7	8	9
(1) Valsad district								
Umargam	Near Umargam	-	-	0	117	412	-	529
	Near of Manekpur	42	33	75	736	587	-	1398
	West of Kala village	28	30	58	20	109	-	187
	Near Kala village	-	-	0	8	948	-	956
		70	63	133	881	2056	0	3070
Navsari	Near Purna village	268	529	797	475	4455	-	5727
	East side of Gulf of Khambhat	-	-	0	-	158	-	158
		268	529	797	475	4613	0	5885
Gandevi	N-E of Ambika river	14	19	33	249	2770	-	3052
	North of Ambika river	14	39	53	70	59	-	182
		28	58	86	319	2829	0	3234
Valsad	Between Ambika river and Par river	-	-	0	382	1437	0	1819
Pardi	South of Par river, Umarsadi village	12	-	12	101	193	0	306
	Kolak	-	-	0	103	188	-	291
		12	0	12	586	1818	0	2416
Total of Valsad district		**378**	**650**	**1028**	**2261**	**11316**	**0**	**14605**

Source: Singh, H.S. (2000).

3

Approach to Valuation of Mangroves

One major reason for the neglect of mangroves in most economies is that its full value is neither calculated nor recognised. This is largely because many of the products and services of mangroves do not enter the market. It is necessary therefore, to compute monetary value of the products and services of mangroves to enable policy makers and people to appreciate its importance in the economy. This will ensure proper growth of mangroves on the one hand (as resources will be allocated to its growth looking to its economic value) and ensure sustainable use of mangroves on the other hand (as mangroves will not be overused in view of its high value).

This chapter, which discusses our approach to valuation of mangroves, is divided into two parts: framework for valuation and methodology of valuation.

Framework for Valuation of Mangroves

A framework for estimating the total economic value of mangroves needs to be multidimensional, as the value needs to be seen in multiple ways:

Replacement Cost versus Development Cost

To start with, the value of mangroves can be seen as preservation/ replacement value as well as developmental value. The preservation value or the replacement cost will refer to the cost that one has to incur to regenerate the lost mangroves. The replacement cost will be computed with reference to any specific period for which the value is being compiled. This cost will generally refer to the cost of getting/ buying mangroves seeds, setting up a nursery, cost of plantation, cost of care & protection till mangroves are matured, cost of gap filling and other related costs. The developmental cost will include the value of mangroves in terms of its use and non-use values. The replacement cost will be much less than the developmental cost and the former will be the floor cost or the lowest value of mangroves. The gap between the two sets of values will indicate that it costs relatively less to regenerate mangroves, but it will cost much more if mangroves are not regenerated and restored.

Use and Non-use Value

The total value of mangroves can be broadly divided into two parts, namely, use value and non-use value. The use value will consist of the value of the goods and services provided by mangroves to the economy, while the non-use value will be the value that will arise from the mere existence of mangroves.

Use Value: The use value of mangroves can be further divided into direct use value and indirect use value, where the direct use value refer to the direct use or direct interactions with mangroves and its services, and indirect use value refers to the value that stems from the indirect support and protection provided to the economic activity and to property by natural functions of mangroves and by its regulatory environmental functions.

Direct use value can be divided into (1) consumptive use value— the value that can be directly consumed and (2) non-consumptive use value—the on-site use value, such as aesthetic value, open space, recreation, spiritual gains etc.

Indirect use value will be non-consumptive use value, which will refer to the use of mangroves in terms of flood control, protection from strong winds and cyclones etc. It includes ecological functions of mangroves.

Non-use Value: The non-use value of mangroves arises from the mere existence of mangroves. This value can be seen as bequest value, existence value, optional value, stewardship value or intrinsic value. Existence Value refers to willingness to pay for preservation of mangroves even though there is no physical access to the resource, now or in the future. They have no intention of using the resource now or in the future. Existence value is thus:

- Cognitive (mental, psychological, subjective perpetual) value

- Independent of current and future use of mangroves, and

- It has no direct on-site interaction between the individual and the resource, and thus, the interaction is non-consumptive, indirect and off-site.

Option value is the value people attach to the environment just to hold or preserve it or have the option of using it in the future. Bequest Value is the value people are willing to pay to preserve the resources for the sake of their future generation, and stewardship value refers to a situation upon people act as stewards of mangroves.

The Table 3.1 presents the different economic value of mangroves.

Table 3.1

Total Economic Value of Mangroves

Use Value

Direct Use Value	Direct Consumptive Use Value	Fodder, fuel wood, timber, medicine, food etc. Industrial input On-site and off-site fishery
	Direct Non-consumptive Use Value	Recreation, tourism etc.
Indirect Use Value	Functional Benefits	Flood Control, protection against strong winds, cyclone, Biodiversity, Shoreline stabilisation, Carbon Sequestration
Option Value	Future Direct and Indirect Values	Biodiversity, Conserved habitats

Non-use Value

Bequest Value	Preservation of Mangroves for future generations
Existence Value	Value from knowledge of continued existence and Cognitive Value

Value of Mangroves at Different Levels

The total value of mangroves can also be seen as the sum total of values manifested at different levels, i.e. at the local or community level, at the state level, at the country level and at the global level.

At the local or community level, mangroves provide fodder, fuel wood, timber etc; supply crabs, mudskippers etc. by way of on-site fishing; protect local communities from saline winds, cyclone etc. At the state level, the benefits are much wider. Mangroves, at the state level, enrich coastal and marine life, promote biodiversity, promote off-site fisheries etc; while at the national level, mangroves can contribute significantly by promoting carbon sequestration as well as biodiversity. Rich growth of mangroves on the coast of Gujarat is likely to have a significant impact on the marine biodiversity to other coastal regions, particularly in neighboring states. Also, carbon sequestration promoted by mangroves has impact on the environment of the entire country. At the global level also, a rich growth of mangroves can make a contribution in reducing global warming. In short, the value of mangroves needs to be seen beyond its local value.

Table 3.2

Value of Mangroves at Different Levels

Levels	Value
Local/Community	Fodder, Fuel wood, timber, good etc. Protection from strong winds, salinity etc. Inshore and offshore fisheries.
State (Gujarat)	Offshore fisheries and marine life. Protection from cyclones. Enriched coastal and marine life, biodiversity.
Country (India)	Enriched biodiversity. Carbon sequestration.
Global level	Carbon sequestration.

Livelihood Protection and Distributional Dimension of Valuation

An important dimension of the value of mangroves is that it protects and promotes livelihood of lakhs of people on the Gujarat coast. Mangroves support the livelihoods of the thousands of households engaged in animal husbandry and fishery, they protect coastal lands from the strong winds and salinity ingress, they meet basic needs like fuel wood, timber etc. and protect the life and property of thousands of households living in the coastal areas. This contribution of mangroves cannot be captured adequately in monetary terms, as it has an important dimension of distribution of benefits. If mangroves are destroyed for an industrial unit, the loss will occur to hundreds of households, who will be deprived of their livelihoods.

It will be necessary to include this dimension while taking any decision about clearing mangroves for any developmental project like industrial project, infrastructural project, etc. For example, it will be necessary to give a high priority to these losers in the distribution of benefits of the developmental project.

Present and Potential Value of Mangroves

Another important dimension of valuation of mangroves is about the wide gap between its present and potential values. This is because the value of mangroves increases multifold when the patchy, stunted and under grown mangroves turn in to matured plants.

Valuation of mangroves needs to be seen in terms of its potential contribution to the economy and any decision about mangroves should be based on its long-term contribution to the economy and not on the basis of its present status. Since the value of mangroves increases at an exponential rate with the improvement in its quality, its value for the coming years should be calculated accordingly. A rich belt of well-grown mangroves can contribute to the economy at all the levels, starting from local level to the global level. It will contribute not only to local community by supplying fodder, fuel wood, timber, food and fisheries etc., but it will contribute significantly at the state economy by promoting its multiple uses like raw material for a number of medicines, industrial products, etc. and by enriching biodiversity and at the country and global level by rich carbon sequestration.

It will be necessary to incorporate this dimension of mangroves while estimating its value in the coming years, say coming 5 to 10 years.

To sum up, the total framework for valuation of mangroves can be presented as follows:

Table 3.3

Framework for Valuation of Mangroves

Level of Valuation	Use Value		Non-use Value		Replacement Value
	Present	*Potential (Coming 5-10 yrs)*	*Present*	*Potential (Coming 5-10 yrs)*	
Community/ Local level	√	√	√	√	√
State level	√	√	√	√	√
Country level	√	√	√	√	-
Global level	√	√	√	√	-

Methods of Valuation of Mangroves

The Replacement Method

The present study will estimate the replacement cost of mangroves in Gujarat. As seen earlier, the replacement cost will estimate the cost of replacement of the lost mangroves in the state during any specified period. That is, it will be the cost that needs to be incurred to maintain the original level of the mangroves that was there in the beginning of the specified period for which the valuation is done.

Two period have been selected for valuation here: Firstly, the value has been calculated for the loss of mangroves during the past three years from 1998–2001, the period for which the latest and comparable data on mangroves are available. The value of the gain or loss in the area under mangroves has been calculated by computing the cost of restoring this area under mangroves. Attention has also been given to the changes in the quality of mangroves that has taken during this period. Secondly, valuation has also been done from the long-term perspective. Here the cost of regeneration of mangroves has been calculated for the potential area under mangroves. Since we do not have any time series data on the status of mangroves in the state, there are no estimates available for the long-term loss of mangroves. Keeping this in mind the replacement cost of mangroves has been calculated for all the potential area identified as suitable for growth of mangroves. This potential area for mangroves, as seen earlier, has been estimated by Singh (1999).

The seven models/sites of the REMAG project have been used for computing the cost of mangroves. The coastal region of the state, as discussed above, can be divided into five regions: (1) The Rann of Kachchh, (2) The Gulf of Kachchh, (3) The Saurashtra coast, (4) The Gulf of Khambhat and (5) The South Gujarat coast.

The area for restoration of mangroves has been estimated for each of the regions separately and the cost of restoration for each region has been calculated using the costs incurred under different REMAG experiments/models. In addition, the different models of mangroves plantation used in the Marine National Park also have been used for the purpose of valuation. This valuation provides full estimates of the cost of restoration of mangroves on the coast of Gujarat, keeping in mind the differences in the costs in the different regions. It provides the estimate of the cost that the government and other stakeholders need to incur to ensure healthy growth of mangroves on the suitable sites on the Gujarat coast.

There are certain underlying assumptions under these calculations that need to be made explicit:

1. It is assumed that the present cum potential area under mangroves is the optimum area for mangroves in the state. This assumption is realistic because as per our present information and knowledge, it is not possible to go beyond these areas for mangrove plantation.

2. It is also assumed that the area suitable for mangroves is available for mangroves plantation. This assumption is slightly difficult, as the state government has allotted coastal lands to several alternative uses, such as ports & jetties, industries, shipping, infrastructure etc. However, it is assumed that the state government will be committed to allot this identified area, which constitutes just 6.67 per cent of the total coastal wetlands for mangroves.[1]

Estimating Direct and Indirect Use Value of Mangroves

An alternative value of mangroves has been calculated using the direct and indirect use value of mangroves. This has been done with

---

1. It has been estimated by experts that the total coastal wetlands in Gujarat is of 2508300 ha. or 25083 sq. km. (Singh, 1999). SAC has estimated this area at 2850000 ha. or 28500 sq. km. As against this, the area under mangroves and the potential area for mangroves come to 167540 ha. or 1675.40 sq. km., which comes to 6.67 per cent of the total coastal wetlands in the state.

the help of primary surveys in the selected villages from each of the coastal regions as well as using the available studies/literature on mangroves. The available studies on mangroves have been used when (1) conducting primary studies on the coast is not likely to help and (2) when the available studies are relevant to Gujarat's coastal regions.

Present Use Value of Mangroves: The present use value of mangroves, based on its present uses by people on the coast, has been calculated with the help of well designed primary surveys in the selected coastal villages.

The selection of villages has been done in a way that (1) they are coastal villages with relatively good growth of mangroves, (2) the villagers depend on mangroves for their different needs and (3) there is a scope for regeneration of mangroves. The selected villages also are the villages where GO/NGOs have undertaken regeneration of mangroves. These villages thus provide the cost of regeneration of mangroves in the different regions in the state. Table 3.4 presents information about the selected villages from the different coastal regions in the state.

Table 3.4

Selected Villages for Valuation

Sr. No.	Regions	Districts	Talukas	Villages
1.	Kachchh and Jamnagar	Kachchh	Jakhau	Asirwandh, Gunao
			Lakhpat	Lukky
		Jamnagar	Jamnagar	MNP area
2.	Saurashtra (for)	Ahmedabad	Dhandhuka	Mahadevpura Bhangadh
3.	Gulf of Khambhat	Ahmedabad	Dhandhuka	Mahadevpura Bhangadh
		Anand	Khambhat	Tada talav
4.	South Gujarat (for)	Bharuch	Hansot	Kantiajal
			Jambusar	Neja, Nada

Source: Primary Survey.

Since no data are available for the Saurashtra region, it is assumed that the cost of regeneration will be similar to the costs in

Dhandhuka, which is close to this region. Similarly, since no regeneration data are available for South Gujarat, it is assumed that the costs of regeneration in this region will be equal to the costs in Hansot *taluka*, which is in some ways similar to South Gujarat coastal region.

In Kachchh-Jamnagar area the regeneration work is of two types, regeneration by the forest department in the protected area of the MNP & S, and regenerations by communities in non-forest areas. Two sets of sites have been selected therefore, to represent the two approaches. The models of the MNP area presents the costs under the forest department while the costs in Asirawandh village presents the costs by communities.

The NGOs engaged in the regeneration of mangroves under the REMAG project provided useful data on the past status of mangroves in the villages as well as the cost of regeneration of mangroves in these regions. They also provided rich data on the villages, the processes used in regeneration, the ground level problems in the regeneration etc. The following are the NGOs engaged in regeneration of mangroves in the selected sites:

Table 3.5

Selected Sites and the NGOs Engaged in Restoration

District/Taluka	Village	NGO
Kachchh/Jakhau	Asirawanth Gunao Lukky	Gujarat Institute of Desert Ecology (Guide)
Ahmedabad/Dhandhuka	Mahadevpura Bhangadh	Mahiti Gram Vikas Snastha
Anand/Khambhat	Tadatalav	Deheda Sangh, Daheda
Bharuch/Jambusar	Neja Nada	Vikas Centre for Development
Bharuch/Hansot	Kantiajal	Vikas Centre for Development

Source: Primary Surveys, 2003.

Estimating Dependences of Sample Villages on Mangrove: In order to estimate the dependence of the selected villages on mangroves for direct and indirect use value, the following schedules were designed:

1. Village Schedule for collecting village level information about the socioeconomic and geophysical characteristics of the villages, the past and present status of mangroves around the village, the village level efforts for regeneration of mangroves etc. The PRAs conducted by NGOs and their progress reports as well as the base line surveys conducted by experts in the selected villages also provided useful village level information.

2. Household Listing Schedule, which covered all the households residing in the selected villages, collected information about the major household characteristics, such as, occupation, land holding, income, household size etc. as well as the dependence of the household on mangroves for fuel wood, fodder, timber, collection of crabs and mudskippers in the water ways, medicines, fisheries etc.

3. Household Sample Survey schedule to collect detailed information about the use of mangroves from the households that use mangroves from different purposes, and

4. Contingency Valuation Schedule for estimating the indirect use value of mangroves to the selected villages, in terms of protection from strong winds, saline winds and salinity ingress in land and water.

The household sample survey schedule, prepared to collect detailed information about the use of mangroves for different purposes, covered the households selected through stratified random sampling. The village households were stratified on the basis of their use and non-use of mangroves for different products and services and their income levels. The sample was then selected using random sampling methods. The schedule was designed to collect information about the pattern of the use of mangroves by the households engaged in different occupations across the different seasons as well as the quantity and price (if paid) of the mangroves used. In order to estimate the contribution of mangroves to fishery, we estimated the household engaged in fisheries, estimated their fishing income, and based on the dose response coefficient between fisheries and mangroves, calculated the contribution of mangroves into the incomes from fishery.

The Contingency Valuation Schedule was designed to estimate the contribution of mangroves in term of protection to agriculture, property, infrastructure, life etc. The four schedules together help in estimating the dependence of the villages on mangroves, i.e. the use value of mangroves to the villages.

Use Value of Mangroves for Potential Areas: In order to estimate the use value of mangroves for the potential area under mangroves, the estimates of the present value have been suitably scaled up for the potential areas.

Estimating Non-use Value of Mangroves

The non-use value of mangroves, i.e. existence value and bequest value, can be compiled using contingency valuation methods. However, the WTP (Willingness To Pay) value cannot be captured from local villagers/population because: (1) it does not occur only to them, but it occurs to the entire state as well as to the country, and to an extent, to the world, and (2) it will be difficult for local people to appreciate the non-use value independently of their use value. Ideally speaking, it will be useful to undertake contingency valuation of mangroves by compiling WTP of people who live in the state or in the country, and who are educated and aware about the non-use value of mangroves.

Since the present study does not have scope for conducting this valuation, some available studies have been used for compiling these Bequest and Existence values.

Potential Value of Mangroves

Though mangroves have a large number of economic uses, it is used for limited purposes in Gujarat. This seems to be because: (1) people are not aware about the large range of uses; (2) they are not interested in exploiting the variety of uses, and/or (3) the present status of mangroves—their growth, their quality and their diversity—does not alter exploitation for many uses. However, a comprehensive valuation of mangroves should incorporate these uses, particularly because, with the expansion and improvement in mangroves, it will be possible to exploit the full potential of mangroves economically.

This study therefore, includes valuation of this variety of additional uses of mangroves. These will include use of mangroves for dyes, medicines, beewax & honey, on-site and off-site fisheries, recreation and tourism, biodiversity benefits, shoreline stabilisation, protection from cyclones, carbon sequestration etc. These values will be compiled using the available literature and studies.

It needs to be underlined that it is important to include these potential uses of mangroves, as these benefits will occur to the state, when the mangroves are well grown. To put it differently, any decision about mangroves in the state must be taken looking to their full potential contribution to the economy.

Another aspect of the potential value of mangroves is the exponential rate at which the benefits from mangroves will grow. The present benefits from mangroves, such as fodder, fuel wood, timber etc. will increase at a much faster rate with the growth of mangroves. While compiling the value of mangroves for the coming 5-10 years, it will be necessary to take note of this phenomenon.

To sum up, the valuation of mangroves in Gujarat in this study will include the following:

- As far as the replacement cost/value of mangroves is concerned, it will include: (1) the replacement value of mangroves for the period between 1998-2001, and (2) the replacement value of mangroves for full restoration of mangroves in the state.

- As far as the valuation of the direct use value of mangroves is concerned, it will include:

  - Direct use value of mangroves with the present coverage of mangroves (for the current year).

  - Direct use value of mangroves with the plantation on potential area under mangroves.

  - Direct use value of mangroves for the coming five years based on the exponential growth of benefits of mangroves, with the present and full coverage of mangroves.

- Indirect Use Value of Mangroves
  - Value for livelihood promotion through fisheries etc.
  - Value for protection of life and property.
  - Value of other ecological functions of mangroves, such as carbon sequestration.
- Non-use Value of Mangroves
  - For carbon sequestration.
  - For coastline stabilisation.
  - For preservation value.

4

Replacement Value of Mangroves

As discussed in the earlier chapter, the replacement cost method estimates the cost of replacement of mangroves in the state. It is defined as the cost that needs to be incurred to maintain the original level of the mangroves that was there in the beginning of the period for which the valuation is done.

Two period have been selected for valuation here: Firstly, the value has been calculated for the loss of mangroves during the past three years from 1998–2001, the period for which the latest data on mangroves are available. The value of the gain or loss in the area under mangroves has been calculated by computing the cost of restoring this area under mangroves. Attention has also been given to the changes in the quality of mangroves that has taken during this period. Secondly, valuation has also been done from the long term perspective. Here the cost of regeneration of mangroves has been calculated for the potential area under mangroves. Since we do not have any time series data on the status of mangroves in the state, there are no estimates available for the long-term loss of mangroves. Keeping this in mind the replacement cost of mangroves has been calculated for all the potential area identified as suitable for growth of mangroves. This potential area for mangroves, as seen earlier, has been estimated by H.S. Singh (1999).

The Sample Villages

The sample villages located in the different coastal region of the state represent the typologies of mangroves in the state.

Asirawandh is a small coastal village, with about 30 households, located near Jakhau in Abdasa *taluka* of Kachchh district. It is typically a *Maldhari* community village whose main occupation is animal husbandry. About 22 households of the village depend on animal husbandry either as their main occupation (18) or as their secondary occupation, and about 13 households depend on fisheries either as their main occupation (7) or as their secondary occupation (6). That is, about 85 per cent of the households in the village depend either on animal husbandry or fisheries, both these occupations depending considerably on mangroves. There are only two farmers in the village with some land. About 18 households depend on casual labour for their livelihood and of these 5 households have unskilled labour as their main occupation. Unskilled labour is performed in nearby saltpans, on farms or in Sanghi Cement factory.

The village does not have any health facility; no doctor or a nurse visits the villages anytime. The nearest dispensary is at Nalia, which is 24 km. away. There is neither any electricity nor any bus services to this village. Till now there was no school in the village. A school has been started by GUIDE about four months ago, but the school has just one room (*kutcha* room) where all the children from 4 to 14 sit during the school hours. People in the villages used to buy drinking water for themselves and for their animals, spending about Rs. 1200–Rs. 2000 per household per month, till recently. Now water is provided to the village under the pipeline scheme. The water stinks (the pipes are stinky) which make it useless for human consumption. Most houses in the village are *kutcha* houses with none having any electricity connection or sanitation facility.

About 92 per cent population of the village is illiterate, with all adult women being illiterate. Some boys and girls, however, have started attending the nearby school. The village is not very poor in the sense that about 8 of the total 30 households have the BPL card (the number of the poor seems to be much less), as most of the households are able to eat enough thanks to their incomes from buffaloes, camels, and fisheries as well as from nearby saltpans or the cement factory.

Gunao and Lukky villages of Lakhpat *taluka* of Kachchh district are small coastal village, with 60 and 28 households respectively, largely depending on animal husbandry and fishery. About 90 per cent of the households of these villages own animals, which mainly depend on mangroves for fodder. Also, about 11 households in Gunao (18.3 per cent) and 4 households in Lukky (7.1 per cent) depend on fishery as their main occupation, though many more households are engaged in fisheries as their secondary occupation. Agriculture is a minor occupation in these villages, with 18 households in Gunao (30 per cent) and 10 households in Lukky (35 per cent) engaged in cultivation. About 21 per cent (Gunao) to 35 per cent (Lukky) households in these villages are engaged in unskilled labour, as their main occupation. They work mainly on farms, saltpans and in nearby industries.

Like Asirawandh, the level of literacy is very low in these villages with 100 per cent illiteracy in Lukky and 60 per cent population being illiterate in Gunao. There is no school in Lukky with no child attending any school. In the case of Gunao, where there is a primary school, about 80 per cent of children attend the school. There is no dispensary or any other health facility in any of these villages. The incidence of poverty is very low in these villages with 8 per cent and 16 per cent households having BPL card in Lukky and Gunao respectively. In short, Gunao and Lukky are also backward villages with poor amenities and facilities, but low incidence of poverty and high incidence of human poverty!

Bhangadh and Mahadevpura are coastal villages in Bhal area and Dhandhuka *taluka* of Ahmedabad district. Both are relatively large villages with 442 households (Bhangadh) and 210 households (Mahadevpura). These villages are multi-caste villages with people belonging to *Darbar* (*Garasiya*) caste, *Patels*, *Brahmin*, as well as the schedules castes and other backward castes. The main occupations of the villages are agriculture and agricultural labour, with about 90 per cent to 95 per cent of the households engaged in these occupations. Of these 63 per cent households in Bhangadh and 79 per cent households in Mahadevpura have agricultural/rural labour as their main occupation. As the agriculture here is mainly rainfed, there is large scale seasonal out migration from these villages to distant places in search of work.

About 30.8 per cent households in Bhangadh and 25.4 per cent households in Mahadevpura own livestocks and most of these households rely on mangroves for fodder for the large part of the year. The villages had good mangroves in the past, which helped them in getting fodder and also fuel wood. Since the mangroves have been badly neglected, the products and services provided by them have declined significantly.

Both the villages have a very high incidence of illiteracy, 65 per cent in Bhangadh and 68 per cent in Mahadevpura. The incidence of illiteracy of women is much higher around 75 per cent in both the villages. Majority of children, 66 per cent in Bhangadh and 76 per cent in Mahadevpura do not attend school. More than 96 per cent of households, in Bhangadh and 18 per cent households in Mahadevpura live in *kutcha* houses and without any individual sanitation facility. The villages have a *pucca* approach roads but poor bus facility. In short, the villages are poor and backward villages.

Tadatalav is a relatively small village of about 111 households and about 600 population. It is located about 3-4 km. away from the sea coast in Khambhat *taluka*. It is 20-25 km. away from Khambhat town. Most of the land in the village has become saline and barren thanks to the advancing sea. Sea water reaches the village during high tide. About 50 per cent land of the village has become saline and useless for cultivation. However, in the absence of alternative occupation, about 43 per cent of the households depend on agriculture and about 50 per cent of the households depend on agricultural or rural labour. That is, about 93 per cent of the households depend on agriculture and unskilled labour. Many families (more than 50 per cent) migrate seasonally to distant places in search of work during the non-*Kharif* seasons.

The village has an all weather approach road, with the bus coming to the village twice a day. About 33 per cent of men and 57 per cent of women in the village are illiterate. There is a primary school in the village where more than 60 per cent of children go regularly. There is no public health service in the village, but a nurse comes once in a month or so. Drinking water sources are local, but not reliable. In short, Tadatalav is an environmentally degraded village where the livelihood of people has been adversely affected by depleted and degraded natural resources.

About 51 per cent of the households own livestocks, but only 8.8 per cent of them depend on mangroves for fodder as mangroves have depleted and degraded badly during the past decades. Tadatalav had good mangroves forest about 30 years ago, when mangroves helped the village to face droughts. People in the village used mangroves as fuel wood and fodder, and sold it to earn a living. Over the past 30 years, however, the mangroves have depleted and degraded, lending to severe shortage of locally available fodder and fuel wood.

Kantiajal is another poor coastal village located in Hansot *taluka* of Bharuch district. It is a small village of about 237 households (1200 population) located about 40 km. away from Ankleshwar. The villages is right on the coast, with the river Kim flowing to the south of the village and river Narmada flowing on the north of the village. The village is located between the two delta regions of the rivers. In 1964, during high tides saline water entered the village and made many farms saline. At present only 100 ha. of the village land is non-saline and good for cultivation, as the large part of the land is under the sea due to the advanced sea. In spite of this, the main occupation of the village is agriculture with 54 per cent of the households engaged in agriculture and 26 per cent of the households engaged in agricultural and rural labour.

The level of literacy is relatively good in this village with 31 per cent population being illiterate. About 75 per cent children are attending the school at present. The village has a *pucca* approach road with a good bus service, electricity connection in 97.5 per cent houses with 75 per cent houses having their own meter. In short, the village is environmentally degraded, but enjoys some amenities and facilities and has a relatively low incidence of poverty.

Fisheries and animal husbandry, both depending on mangroves, are the most important occupations (after agriculture) in the village. About 18 per cent of the households depend on fisheries and 42 per cent households have livestock. Most of the livestock owning households (78 per cent) depend on mangroves for fodder. Thus, about 60 per cent of the households depend on mangroves, to a small or large extent, for their livelihood. The depletion and degradation of mangroves in the recent decades have affected adversely the livelihood of these households.

Neja and Nada are coastal villages located right on the coast in Jambusar *taluka* of Bharuch district. This *taluka* is highly degraded environmentally with salinity ingress in its land and water resources. Neja is a small village with about 98 households while Nada is a much bigger village with 401 households and more than 3000 population.

Nada is located on the sea coast with about 1000 ha. coastal area and is about 32 kms away from Jambusar. River Dhadhar flows to the south of the village and merges into the sea. Saltpans and ONGC oil wells and drilling stations are located close to the village. Nada had thick mangroves forest in the past and still has some good patches of mangroves. About 63 per cent of the households in the village are engaged in agriculture while 20 per cent are engaged in labour work on saltpans and other economic activities. About 2 per cent of the households are engaged in government and private services in and around the village. The village depends on mangroves for animal husbandry and fishery. About 13 per cent of the households are engaged in fisheries and about 65 per cent households have livestock, with 94 per cent of them depending on mangroves for fodder. The mangroves in the village have depleted and degraded badly over past 3-4 decades. Villagers are keenly interested in their revival.

About 68 per cent of the adult population in Nada is literate. The village is relatively better educated. The village has a *pucca* road and with good bus service and has electricity connection in more than 91 per cent of its households (of which only 62 per cent have meters). The village has elementary education facility—with public and a few private dispensaries. The village, however, has problems of adequate employment, which pushes several households out on seasonal migration. In short, this village needs environmental regeneration to promote the livelihood of its people. Regeneration of mangroves will help the village considerably in reviving the livelihood of people.

Neja is a small village on the seacoast. The village has been degraded environmentally due to salinity ingress in land and water. The main occupation of the village is agriculture, with 42 per cent households engaged in cultivation and 50 per cent households engaged in agricultural and rural labour. About 66 per cent households have livestock and 30 per cent of these households depend on mangroves for fodder, 8 per cent households are engaged in fishery and 2 per cent households are engaged in government or private services.

Mangroves are important to the village for the livelihood of people, as about 40 per cent of the households engaged in fisheries and animal husbandry (using mangroves for fodder), depend on mangroves. The status of mangroves at present is very poor, with their stunted growth and scattered patches. However, villagers are keen to develop them for their livelihood and for protection from winds and salinity.

About 65 per cent of the adult population of the village is literate, with the level of literacy higher for men (74 per cent) than women (53 per cent). More than 70 per cent of children are attending school at present. All the households in Neja have electricity connection to their homes though only 60 per cent have legal connection. The village has a *pucca* road and regular bus service. In short, the village has relatively good amenities and facilities.

To sum up, the selected villages differ considerably in terms of their amenities and facilities, non-agricultural employment avenues, and their poverty. However, in spite of these variations, they all have some important common features: (1) they suffer from degraded

Figure 4.1

Village-wise Occupational Structure (Main)

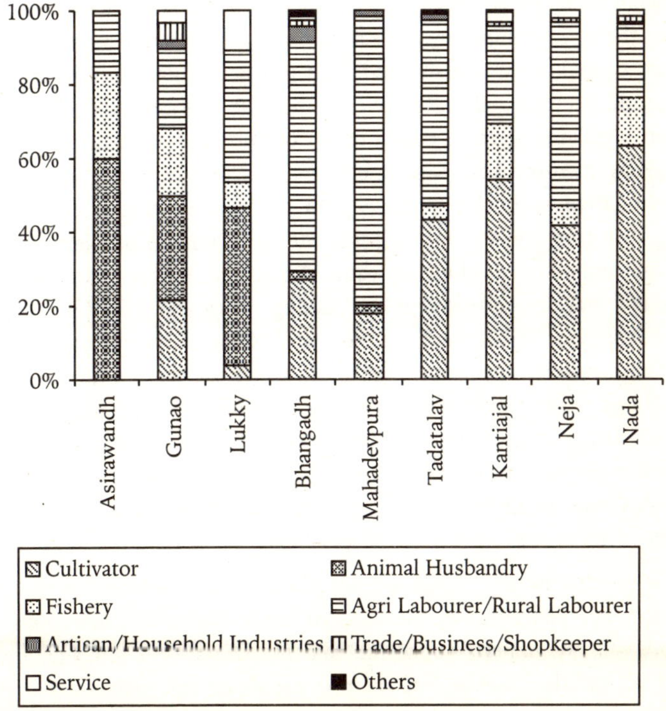

Source: Based on Primary Survey.

coastal environment; (2) they depend on mangroves for livelihood to a significant extent, and (3) ecological regeneration including regeneration of mangroves will help considerably in improving life and livelihood of their population.

Regeneration of Mangroves in the Selected Villages

As mentioned earlier, some NGOs are engaged in the regeneration and restoration of mangroves in the selected villages. Each village is expected to get about 1000 ha. for the purpose. The study used the costs incurred by the NGOs for regeneration of mangroves for the purpose of computing replacement costs of mangroves in the state.

Our visits to the sites and discussion with the NGOs indicated that the NGOs have suffered several problems while undertaking the work of plantation of mangroves. These problems are relating to land acquisition for plantation, nursery making—site, time and protection of nursery, technical support for plantation, protection of plants etc. Though each village was supposed to get 1000 ha. for plantation, even after the official support, the sanctioned land was not made available. Nursery making was also a difficult job, particularly with respect to location, timings and protection. In the case of Asirawandh village the main problems were of poor quality of collected seeds and wash out of seeds from the poly bags due to high current of water, while in the case of Bhangadh and Mahadevpura the success rate at the nursery was very low due to high salinity and temperature problems as well as destruction of seedlings by camels. The site, which was selected for nursery, was not getting regular flushing of sea water as it was located at a relatively higher land. At Neja and Nada the major problem was the high salinity and heat. High tidal waves too caused destruction of seedlings. These problems were faced during and after the plantation of seedlings. Grazing by camel after plantation was a common problem in the case of Asirawandh, Mahadevpura, and Bhangadh. The survival rate at none of the sites was more than 60 per cent and in some sites it was as low as 20 per cent.

It seems that the NGOs are not always able to follow the technical advice. For example, in the case of Neja, Nada, Mahadevpura and Bhangadh, the experts suggested that no plantation should take place without physical works, like digging canals on the plantation sites to serve as a natural creek, to facilitate the water flows. However, plantation was done without undertaking physical works. The reason given was the lack of funds for the works. Sometimes there was communication gap between the experts, NGOs and villagers. For example, in the case of Nada, plantation was done in the scorching heat without proper consultation with the experts, with the result that very few seedlings actually survived.

The dedication of the NGOs, however, is worth noting. The NGOs also have been successful in building strong community organisations.

Replacement and Restoration Cost

Replacement and restoration value of mangroves has been calculated for two periods: (1) for the period between 1998 and 2001 and (2) the long term replacement value under the assumption that all the potential areas for mangroves was under mangroves a few decades ago.

Replacement Value for 1998–2001: Table 2.6 (chap. 2, p.49) presents data on the coverage of mangroves in 1992, 1998, 1999 and 2001 as per the alternative estimates available. The SAC estimates (1992) are available for the scale of 1: 1,25,000 and are not comparable either with the FSI estimates (2001) or the GEER foundation estimates (1998) both of which are for the scale of 1: 50,000. The FSI estimates of 1999 and 2001 also are not comparable as the 1999 estimates are for the scale 1: 1,25,000 while the 2001 estimates are for the scale of 1: 50,000. The estimates of GEER Foundation for the year 1998 and the FSI estimates of 2001 are however, comparable as both are in the scale of 1: 50,000. The replacement cost for mangroves therefore, has been calculated using these data. Both the sources provide data on mangroves under dense forests and under sparse forests. Both the sets of data have been used for calculating the replacement cost. Table 2.6 shows that the dense mangroves cover has declined from 455.42 sq. km. in 1998 to 182 sq. km. in 2001, i.e. it has declined by 182 sq. km. during the period. The sparse forest on the other hand has increased from 482.7 sq. km. in 1998 to 728 sq. km. in 2001 implying an increase of 243.3 sq. km. during the period. The total area under mangroves has declined from 938.12 sq. km. to 908 sq. km. by 29.12 sq. km. during the period (Table 2.6).

Table 2.6 shows that the dense mangroves area has declined considerably in both Kachchh and Jamnagar area, by about 50 sq. km.

and by 226.18 sq. km. respectively during 1998–2001. The area under dense mangroves has also declined in Surat, Valsad and Rajkot, while it has increased marginally in Ahmedabad, Bhavnagar and Junagadh and significantly in Bharuch, (from 5.69 sq. km. in 1998 to 17.00 sq. km. in 2001). The decline in dense forest may be mainly due to natural causes for which no clear explanations are available at present. The main reason seems to be the lack of protection, as most of the mangroves in the state (except for those in Jamnagar and Kachchh districts) are outside the preview of the forest department. These mangroves are protected from natural forces nor from indiscriminate cutting. The increase on the other hand appears to mainly due to new plantation or due to degraded dense mangrove turning into sparse mangroves. The per hectare cost of regeneration of dense mangroves in each region has been calculated using the different NGO experiments of regeneration. These costs of regeneration are presented in Table 4.1. The costs are also presented in the Figure 4.2.

Table 4.1

Cost of Regeneration per Hectare

Type of Cost	As per NGO Models					As per MNP Models		
	Asirawandh	Mahadevpura	Neja	Nada	Kantiyajal	SLM-I	SLM-II	SLM-III
Capital Cost of Plantation	3988	4230	7609	2513	6948	4841	2568	6064
Gap Filling Cost After One Year	3584	2500	909	734	288	914	564	1092
Cost of Protection (For 5 Years)	9000	9000	9000	9000	9000	2520	2520	2520
Total cost	16572	15730	17518	12247	16236	8275	5652	9676

Source: Estimated from the actual expense on regenaration at different sites in 2003-04.

Table 4.1 shows that the costs are not the same in the different locations. This is mainly because the physiographic conditions are different in different region, i.e. the climate, salinity, tidal flows, wind velocity, substrata soil etc. are different in different region. The other reasons are: (1) the seeds are not always available locally, with the result that in some regions seeds are brought from distance raising the cost of seeds; (2) the methods adopted for plantation are different, as in some regions direct plantation is possible while in other regions plant nursery is a necessity, which raises the cost of plantation; (3) the rates of success are different in different regions, resulting in the differences in the costs of gap feeling, and (4) the need for protection which differs from region to region, depending on the local situation. Using the data on the cost of plantation in the different regions of the state, the value of the changing area under mangroves has been calculated in Table 4.2. As the table shows, the value of the loss of dense mangroves in the state during 1998-2001 comes to Rs. 40.71 crore.

Figure 4.2

Site-wise Regeneration Cost (Rs. per Ha.)

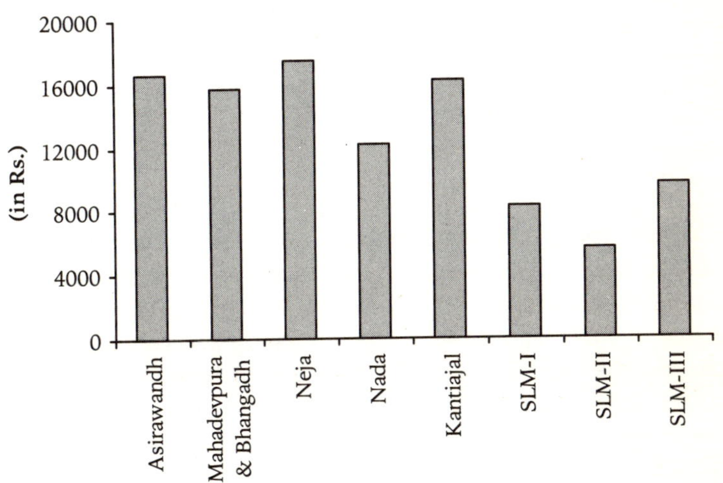

Source: Based on Primary Survey.

Similarly, the value of the increase in sparse forests has been calculated under the assumption that the per hectare cost of regeneration of sparse mangroves is half of the cost of regeneration of dense mangroves.

Table 4.2

Valuation of Change in the Cover of Dense Mangroves (1998-2001)

District	Dense Mangroves Cover in 1998 (sq. km.)*	Dense Mangroves Cover in 2001 (sq. km.)**	Change (in sq. km).	Change (in Hectare)	Cost of Regeneration of Dense Mangroves (in Rs. Cr.)
Ahmedabad	0.09	1	0.91	91	0.14
Bharuch	5.69	17	11.31	1131	1.73
Bhavnagar	9.19	10	0.81	81	0.13
Jamnagar	78.65	28	-50.65	-5065	-3.99
Junagadh	0.65	1	0.35	35	0.06
Kheda	0	0	0	0	0.00
Kachchh	344.19	118	-226.19	-22619	-37.48
Rajkot	0.98	0	-0.98	-98	-0.16
Surat	12.2	8	-4.2	-420	-0.68
Valsad	3.78	1	-2.78	-278	-0.45
Total	455.42	182	-271.42	-27142	-40.71

Note: Negative (-ve) signs indicate the cost to be incurred for regeneration where as Positive (+ve) signs indicate the gain due to natural increase in area under mangroves.

Source: *: Singh, H.S. (2000).

　　**: FSI, *State of Forest Report 2001.*

Table 4.3 shows that the increase in sparse forests has taken place mainly in Jamnagar and Kachchh. In Jamnagar the increase has been from 62.79 sq. km. in 1998 to 114 sq. km. in 2001 implying an increase of 51.21 sq. km. while the increase in Kachchh district has been of 204.82 sq. km. from 383.18 sq. km. in 1998 to 588 sq. km. in 2001. Rajkot, Surat and Valsad have experienced a sharp decline even in sparse mangroves during the period. Junagadh also has experienced a sharp decline in sparse mangroves from 10.35 sq. km. in 1998 to almost zero in 2001. Ahmedabad, Bharuch and Bhavnagar, however, have experienced a marginal increase in the area under sparse mangroves during the period. Table 4.3 presents the cost of the changes in the area under sparse mangroves. It shows that there has been an increase of about Rs. 17.4 crores in the value of sparse mangroves during 1998-2001.

Table 4.3

Valuation of Change in the Cover of Sparse Mangroves (1998-2001)

District	Sparse Mangroves Cover in 1998 (sq. km.)*	Sparse Mangroves Cover in 2001 (sq. km.)**	Change in sq. km.	Change in ha.	Cost of Regeneration of Sparse Mangroves (in Crs. of Rs.)
Ahmedabad	0.7	1	0.3	30	0.02
Bharuch	8.99	11	2.01	201	0.15
Bhavnagar	5.96	6	0.04	4	0.00
Jamnagar	62.79	114	51.21	5121	2.01
Junagadh	10.4	0	-10.4	-1040	-0.79
Kheda	0	0	0	0	0.00
Kachchh	383.18	588	204.82	20482	16.97
Rajkot	4.59	1	-3.59	-359	-0.14
Surat	9.64	5	-4.64	-464	-0.38
Valsad	6.5	1	-5.5	-550	-0.45
Total	482.7	726	234.6	23460	17.41

Source: *- Singh, H.S., (2000).

**- FSI, *State of Forest Report 2001.*

Tables 4.2 and 4.3 together show that the state has lost Rs. 23.3 crores worth of mangrove resources during the period 1998-2001. The value of the loss per year comes to about Rs. 7.8 crores. To put it differently, by spending Rs. 23.3 crores on regeneration of mangroves, the state can make up for the loss of mangroves experienced during 1998-2001.

Value of Full Restoration of Mangroves in the State

The long-term replacement value or full regeneration value of mangroves has been computed under the assumption that the mangroves in the past covered its full potential area. The long-term loss of the area for mangroves has been presented in the Tables 4.4 and 4.5

Table 4.4

Cost of Full Regeneration

(in Crores of Rs.)

District	Potential Area of Mangrove (Ha)	Model Used	Plantation Cost	Gap Filling Cost	Maintenance Cost for Five Years	Total Cost of Regeneration
Kachchh	32734	Asirawandh	13.1	11.7	29.5	54.3
Rajkot	2827	Asirawandh	1.1	1.0	2.5	4.7
Jamnagar	11146	Salt Lease Models	5.0	0.7	2.8	8.5
Junagadh	40	Bhangadh & Mahadevpura	0.0	0.0	0.0	0.1
Bhavnagar	1999	Bhangadh & Mahadevpura	0.8	0.5	1.8	3.1
Ahmedabad	5253	Bhangadh & Mahadevpura	2.2	1.3	4.7	8.3
Anand	1624	Neja	1.2	0.1	1.5	2.8
Bharuch	3729	Neja, Nada, Kantiajal	2.1	0.2	3.4	5.7
Surat	2102	Kantiajal	1.5	0.1	1.9	3.4
Valsad	2261	Kantiajal	1.6	0.1	2.0	3.7
Total	63715		28.7	15.8	50.1	94.6

Source: Primary Survey, 2003.

Table 4.5

Total Cost of Regeneration

(in Crores of Rs.)

District	Cost of Regeneration in Potential Area	Cost of Regeneration of Dense Mangroves Area during 1998-2001	Cost of Regeneration of Sparse Mangroves Area during 1998-2001	Total Cost of Regeneration of Mangroves
Ahmedabad	-8.26	+0.14	+0.02	-8.10
Bharuch	-5.72	+1.73	+0.15	-3.83
Bhavnagar	-3.14	+0.13	+0.00	-3.01
Jamnagar	-8.53	-3.99	+2.01	-10.50
Junagadh	-0.06	+0.06	-0.79	-0.80
Kheda	-2.84	+0.00	0.00	-2.84
Kachchh	-54.25	-37.48	+16.97	-74.76
Rajkot	-4.69	-0.16	-0.14	-4.99
Surat	-3.41	-0.68	-0.38	-4.47
Valsad	-3.67	-0.45	-0.45	-4.57
Total	-94.58	-40.71	+17.41	-117.88

Note: Negative (-ve) signs indicate the cost to be incurred for regeneration where as Positive (+ve) signs indicate the gain due to natural increase in area under mangroves.

Source: Primary Survey, 2003

The potential area has been taken from the estimates made by H.S. Singh for the year 1998 (1999). The cost of regeneration has been calculated region wise, using the same costs per ha. as in the above exercise. The total value of restoration comes to Rs. 94.6 crores. Since the total restoration cost would include the cost of regeneration in potential area as well as the cost of regeneration of the depletion and degradation mangroves during 1998-2001, the total cost of regeneration of mangroves for the state will be Rs. 117.8 crores (Table 4.5). In other words, the state can regenerate mangroves on all the possible coastal area, which includes the present area as well as the potential area, by spending a relatively small amount of Rs. 117.8 crores. The regeneration will of course take about five years.

Annexure A-4.1

Occupational Pattern in the Selected Villages

A. Main Occupation

Village	No of HH			% of HH with Main Occupation as						
		Farmer	Agri Labourer/ Rural Labourer	Animal Husbandry	Fishery	Artisan/ Household Industries	Trade/ Business/ Shopkeeper	Service (Govt.)	Service (Others)	Others
Asirawandh	30	0	16.7	60.0	23.3	0	0	0	0	0
Gunao	60	21.7	21.7	28.3	18.3	1.7	5.0	0.0	3.3	0.0
Lukky	28	3.6	35.7	42.9	7.1	0.0	0.0	3.6	7.1	0.0
Bhangadh	442	26.9	62.4	2.0	0.0	4.3	1.8	0.7	0.0	1.8
Mahadevpura	210	18.1	78.6	1.9	0.0	1.4	0.0	0.0	0.0	0.0
Tadatalav	111	43.2	50.5	0.0	3.6	1.8	0.0	0.0	0.0	0.9
Kantiajal	237	54.0	26.2	0.0	15.2	0.4	0.8	2.1	0.4	0.8
Neja	98	41.8	50.0	0.0	5.1	0.0	1.0	0.0	2.0	0.0
Nada	401	63.1	20.0	0.0	13.0	0.7	1.5	0.5	1.0	0.2

B. Secondary Occupation

Village	No of HH	% of HH with Secondary Occupation as								
		Farmer	Agri Labourer/ Rural Labourer	Animal Husbandry	Fishery	Artisan/ Household Industries	Trade/ Business/ Shopkeeper	Service (Govt.)	Service (Others)	Others
Asirawandh	30		43.3	13.3	20.0				0.0	0.0
Gunao	60	21.7	11.7	18.3	10.0				0.0	0.0
Lukky	28	14.3	14.3	10.7	14.3				0.0	0.0
Bhangadh	442	1.1	39.4	2.7	0.2	5.7	1.6		1.8	2.3
Mahadevpura	210		14.3	4.8		0.5	1.0		0.0	0.0
Tadatalav	111	2.7	24.3	18.9	0.9	10.8	0.0		0.9	0.9
Kantiajal	237	0.4	30.4	10.5	2.5	1.7	1.3	0.8	0.8	0.4
Neja	98	1.0				0.0	0.0		0.0	0.0
Nada	401	2.7	55.9	3.7	1.0	0.5	3.2	0.2	0.2	0.5

Source: Primary Survey, 2003-04.

Annexure A-4.2

Farmers with Different Landholdings in the Selected Villages

Villages	Total HH	Land-less	Number of Farmers	Marginal Farmer (<1 Ha)		Small Farmer (1-2 Ha)		Semi-Medium Farmer (2-4 Ha)		Medium Farmer (4-10 Ha)		Large Farmer (>10 Ha)	
				No	%	No	%	No	%	No	%	No	%
Asirawandh	30	28	2	0	0	1	50	0	0	1	50	0	0
Gunao	60	42	18	1	6	4	22	11	61	2	11	0	0
Lukky	28	18	10	2	20	0	0	3	30	4	40	1	10
Bhangadh	442	296	146	22	15	36	25	50	34	33	23	5	3
Mahadevpura	213	158	55	4	7	3	5	12	22	26	47	10	18
Tadatalav	111	60	51	21	41	14	27	12	24	3	6	1	2
Kantiajal	238	88	150	84	56	42	28	19	13	5	3	0	0
Neja	98	52	46	13	28	7	15	13	28	11	24	2	4
Nada	401	182	219	8	4	37	17	120	55	52	24	2	1

Source: Primary Survey, 2003-04.

Annexure A-4.3

Causes of Degradation of Mangroves as Perceived by Villagers

Villages	Camel Grazing	Grazing of Other Animals	Sea Floor Spreading	Cyclone	Cutting and Lopping	Industries and Saltpans
Asirawandh	0	3.3	0.0	26.7	3.3	46.7
Gunao	1.7	1.7	1.7	60.0	4.9	86.7
Lukky	0.0	3.6	10.7	67.9	3.6	53.6
Bhangadh	96.6	2.0	52.7	0.0	2.0	8.1
Mahadevpura	99.5	0.0	43.2	0.0	1.5	1.4
Tadatalav	9.0	11.7	93.7	0.0	0.9	14.4
Kantiajal	0.0	0.8	53.6	0.4	0.4	0.8
Neja	76.5	18.4	49.0	0.0	3.6	4.1
Nada	79.0	2.7	18.0	0.2	14.0	28.4

The "% of Users with opinion" spans all causes columns.

Source: Primary Survey, 2003-04.

Annexure A-4.4

Educational Status in the Selected Villages

(in per cent)

Education	Asirawandh			Gunao			Lukky			Bhangadh		
	Male	Female	Total	Male	Female	Total	Male	Female	Total	Male	Female	Total
Illiterate	89.7	94.7	92.2	58.5	62.1	60.2	100.0	100.0	100.0	57.6	73.8	64.9
Primary	7.7	5.3	6.5	41.5	37.9	39.8	0.0	0.0	0.0	25.0	14.0	20.0
Secondary	2.6	0.0	1.3	0.0	0.0	0.0	0.0	0.0	0.0	10.4	8.0	9.3
H.S.C	0.0	0.0	0.0	0.0	0.0	0.0	0.0	0.0	0.0	1.7	2.3	2.0
Graduation	0.0	0.0	0.0	0.0	0.0	0.0	0.0	0.0	0.0	5.3	1.8	3.7
Masters	0.0	0.0	0.0	0.0	0.0	0.0	0.0	0.0	0.0	0.0	0.0	0.0
Total	100	100	100	100	100	100	100	100	100	100	100	100

(Contd. ...)

(...contd. ...)

(in per cent)

Education	Mahadevpura			Tadatalav			Kantiajal			Neja			Nada		
	Male	Female	Total	Male	Female	Total	Male	Female	Total	Male	Female	Total	Male	Female	Total
Illiterate	61.3	75.0	67.4	33.3	57.1	44.8	28.8	35.0	31.6	26.2	46.9	35.1	22.7	44.6	32.6
Primary	21.4	17.1	19.5	40.0	28.6	34.5	27.4	34.2	30.5	28.6	12.5	21.6	39.0	40.6	39.7
Secondary	14.5	7.1	11.2	20.0	14.3	17.2	42.5	28.3	36.1	33.3	40.6	36.5	32.9	13.8	24.2
H.S.C	2.3	0.7	1.6	6.7	0.0	3.4	1.4	2.5	1.9	9.5	0.0	5.4	5.5	1.0	3.5
Graduation	0.6	0.0	0.3	0.0	0.0	0.0	0.0	0.0	0.0	2.4	0.0	1.4	0.0	0.0	0.0
Masters	0.0	0.0	0.0	0.0	0.0	0.0	0.0	0.0	0.0	0.0	0.0	0.0	0.0	0.0	0.0
Total	100	100	100	100	100	100	100	100	100	100	100	100	100	100	100

Source: Primary Survey, 2003-04.

5

The Present Direct and Indirect Use Value of Mangroves in Gujarat

In this chapter, the present use value of the products and services provided by mangroves has been calculated for Gujarat. Since the total value can be divided into two parts, direct and indirect use value, this exercise has been divided into two parts with the first part estimating the direct use value and the second part estimating the indirect use value.

The Direct Use Value of Mangroves

The direct use value of mangroves will include the value of mangroves for: (1) fodder, (2) fuel wood, (3) timber for constructing fishing boat, roof etc., (4) vegetable, (5) medicines and other products and (6) collection of mudskippers, crabs and small fish in waterways of mangroves. This valuation has been done using the results of the household surveys conducted in the selected village.

Value of Mangroves as Fodder

An important use of mangroves to the coastal population in the state is as fodder. About 50 per cent of the households living in the selected coastal villages own animals and about 82 per cent of these owners use mangroves as fodder (Table 5.1).

As the above table shows, the per centage of the households owning livestock is relatively large in Kachchh villages (up to 90 per cent) and in Jambusar (more than 65 per cent). However, the dependence of live stock owning households on mangroves for fodder is large in all the villages except Tadalalav (8.8 per cent) and Neja (30.6 per cent). In Kachchh about 92–98 per cent livestock owning households use mangroves as fodder, in Bhangadh and Mahadevpura

Table 5.1

Dependence of Selected Villages on Mangroves

Village	% of Household Having Livestocks	Total Livestock in the Village	% Owners of Livestock Using Mangroves as Fodder
Asirwandh	76.7	266	91.3
Gunao	80.0	376	97.9
Lukky	89.3	210	96.0
Bhangadh	30.8	763	95.6
Mahadevpura	25.4	419	100.0
Tadatalav	51.4	244	8.8
Kantiajal	42.0	271	78.0
Neja	66.3	160	30.6
Nada	65.8	629	94.7
Total	47.6	3338	81.6

Source: Primary Survey, 2003.

the per centage goes up to 100 per cent, in Nada it is 95 per cent while in Kantiajal it is 78 per cent. The number of livestock also is quite large in these villages, varying from 160 in Neja to 629 in Nada and 763 in Bhangadh.

People prefer mangroves for fodder firstly, because these are easily available and are free; secondly, because it is a good fodder that improves milk production and thirdly, because animals like it (Table 5.2).

Of the four major alternative fodders available in the villages, the use of mangroves is the highest during the summer and winter seasons. The maximum number of households using mangroves in these two seasons, with their percentages ranging between 95 per cent to 100 per cent in most villages. The next important fodder is grass, followed by *Khar*, *Bhusu* and *Kadab*. In the rainy season mangroves become inaccessible, turning people to other fodder sources.

Table 5.2

Advantages of Mangroves for Fodder as Perceived by Users

Villages	Reasons for Using Mangrove for Fodder (%)					
	Easy Availability	Free of Cost	Good Fodder	Animals Like it	Other Advantages*	No Special Advantage
Asirawandh	76.2	90.5	100.0	76.2	0.0	0.0
Gunao	70.2	76.6	57.4	44.7	29.8	0.0
Lukky	69.0	79.2	75.0	45.8	4.2	0.0
Bhangadh	62.3	60.0	73.1	61.5	70.8	10.8
Mahadevpura	44.4	46.3	55.6	46.3	59.3	7.4
Tadatalav	60.0	100.0	80.0	40.0	40.0	20.0
Kantiajal	61.5	74.4	69.2	65.4	24.4	24.4
Neja	58.0	80.0	54.0	73.0	24.0	3.0
Nada	36.0	82.4	62.4	62.0	64.0	0.0
Total	54.00	74.00	68.00	61.00	51.00	6.00

Note: *– Other advantages as per the villagers is that animals remain healthy when they eat mangroves.

Source: Primary Survey, 2003-04.

The actual consumption of fodder per household per year has been presented in Table 5.4. The table shows that the annual consumption of mangroves per household (for the sample households using mangroves for fodder) varies from 282 kg. in Tadatalav to 6118 kg. in Kantiajal. When these averages are multiplied with the corresponding total numbers of households using mangroves for fodder in these villages. The total consumption for the villages comes to 3180588 kg. The monetary value of the fodder has been calculated using the local price of fodder in the villages. This price varies from village to village, or from region to region. Using these prices, the money value of mangroves fodder comes to Rs. 1.46 crore.

Table 5.3

Seasonal Consumption of Mangroves as Fodder in Selected Villages

Villages	Sample Household	% HH using as Fodder in Summer				
		Mangrove	Grass	Khar	Bhusu	Kadab
Asirawandh	14	92.9	57.1	64.3	78.6	0.0
Gunao	25	96.0	96.0	96.0	96.0	0.0
Lukky	12	83.3	100.0	8.3	100.0	0.0
Bhangadh	130	46.9	26.2	6.2	0.0	28.5
Mahadevpura	45	62.2	17.8	20.0	0.0	37.8
Tadatalav	5	100.0	0.0	100.0	60.0	40.0
Kantiajal	40	95.0	45.0	2.5	32.5	0.0
Neja	15	100.0	73.3	20.0	40.0	0.0
Nada	116	83.6	27.6	11.2	14.7	3.4
Total	402	72.0	37.0	18.0	21.0	15.0

Villages	Sample Household	% HH using as Fodder in Monsoon				
		Mangrove	Grass	Khar	Bhusu	Kadab
Asirawandh	14	0.0	57.1	64.3	78.6	0.0
Gunao	25	0.0	80.0	96.0	96.0	0.0
Lukky	12	0.0	100.0	8.3	25.0	0.0
Bhangadh	130	0.0	44.6	1.5	0.0	26.9
Mahadevpura	45	0.0	62.2	0.0	0.0	37.8
Tadatalav	5	0.0	100.0	20.0	60.0	40.0
Kantiajal	40	0.0	92.5	2.5	32.5	0.0
Neja	15	0.0	86.7	13.3	33.3	0.0
Nada	116	0.0	82.8	11.2	10.3	5.2
Total	402	0.0	69.0	13.0	18.0	15.0

Villages	Sample Household	% HH using as Fodder in Winter				
		Mangrove	Grass	Khar	Bhusu	Kadab
Asirawandh	14	92.9	57.1	64.3	78.6	0.0
Gunao	25	96.0	96.0	96.0	96.0	0.0
Lukky	12	66.7	100.0	8.3	100.0	0.0
Bhangad	130	41.5	44.6	6.2	0.0	27.7
Mahadevpura	45	62.2	60.0	20.0	0.0	33.3
Tadatalao	5	100.0	100.0	20.0	60.0	40.0
Kantiyajal	40	95.0	95.0	2.5	32.5	0.0
Neja	15	46.7	100.0	13.3	40.0	33.3
Nada	116	81.9	77.6	13.8	13.8	6.9
Total	402	68.0	69.0	18.0	21.0	16.0

Source: Primary Survey, 2003-04.

Table 5.4

Annual Consumption and Value of Mangrove Used as Fodder by Villagers

Villages (Name)	Sample HH (No.)	Total Use ('000 Kg.)	Average Use per HH (Kg.)	HH Using Mangroves (No.)	Total Consumption ('000 Kg.)	Market Value (Rs./ Kg.)	Total Value (Lakhs of Rs.)
Asirawandh	14	49.2	3517	29	102.0	3	3.06
Gunao	25	8.5	340	48	16.3	3	0.49
Lukky	12	6.0	501	24	12.0	3	0.36
Bhangadh	130	340.3	2617	288	753.8	4	30.15
Mahadevpura	45	154.6	3434	97	333.1	4	13.33
Tadatalav	5	1.4	282	5	1.4	4	0.06
Kantiajal	40	244.7	6118	78	477.2	5	23.86
Neja	15	56.4	3762	59	222.0	5	11.10
Nada	116	579.0	4991	253	1262.7	5	63.14
Total	402	1440.1	3582.3	881	3180.6	–	145.54

Source: Primary Survey, 2003.

When this value is scaled up for the total area under mangroves in the state at present, it comes to Rs. 82.76 crores (Table 5.5). This is the current value of mangroves used for fodder in the state for the year 2003-04.

Table 5.5

Value of Mangroves as Fodder

Regions	Value of Mangroves as Fodder per ha. of Mangrove Cover**	Total Area under Mangroves in the Region (ha.)*	Total Value for the Region for the Year
Kachchh and Jamnagar	8689	84900	737735720
Saurashtra	8689	1700	14772093
Gulf of Khambhat	13114	3000	39341372
South Gujarat	23860	1500	35790300
Total	54352	91100	827639485

Sources: *- Forest Survey of India, 2001.
　　　　**- Estimated from Primary Survey, 2003.

Collection of Mangroves: It will not be out of place here to discuss how people collect mangroves. Collection of mangroves is not an easy task, as the area where mangroves grow is muddy and watery. Usually it takes about half an hour to one hour for the villagers to reach the mangroves sites by foot, as they cannot take any bicycle or motorised vehicle to those sites due to the problem of accessibility. The collection usually starts in the morning and continues for 4-5 hours only, as it is difficult to work for longer hours in muddy waters as well as its difficult to carry much of load for such a long distance. Both men and women are engaged in collection of mangroves. Frequently children also participate in this task.

Table 5.6

Who Collects Mangroves?

Villages	No. of Collectors (In sample HH)	Sex-wise				Age Group-wise					
		Male		Female		0 - 14		15 - 60		>60	
		No	%	No	%	No	%	No	%	No	%
Asirawandh	19	2	10.5	17	89.5	2	10.5	17	89.5		0
Gunao	34	13	38.2	21	61.8	3	8.8	31	91.2		0
Lukky	13	10	76.9	3	23.1		0.0	12	92.3	1	7.7
Bhangadh	197	106	53.8	91	46.2	10	5.1	182	92.4	4	2.0
Mahadevpura	62	36	58.1	26	41.9		0.0	62	100.0		0.0
Tadatalav	5	4	80.0	1	20.0		0.0	5	100.0		0.0
Kantiajal	62	33	53.2	29	46.8	8	12.9	52	83.9	2	3.2
Neja	19	9	47.4	10	52.6		0.0	19	100.0		0.0
Nada	172	71	41.3	101	58.7	1	0.6	166	96.5	5	2.9
Total	583	285	48.9	298	51.1	24	4.1	546	93.7	12	2.1

Source: Primary Survey, 2003.

Usually there is no practice of hiring workers for collection of mangroves. However, in some cases workers are hired and paid wages. Table 5.6 shows that both men and women collect mangroves in the villages. Children do not participate in this work in all the villages. Similarly, old people, above 60 years, also participate in small numbers in this strenuous activity.

Table 5.7

Value of Man-days Used for Mangrove Collection

Villages	No. of Sample HH	Man Days Used	Wage Rate (Rs./day)	Total Value of Man-days	Value per HH for the Year (Rs.)	Total No. of HH Collecting Mangroves	Total Value of Collection for the Year (Rs.)
Asirawandh	14	4010	50.0	200500	14321	29	415321
Gunao	25	8040	50.0	402000	16080	48	771840
Lukky	12	785	50.0	39250	3271	24	78500
Bhangadh	130	30180	50.0	1509000	11608	288	3343015
Mahadevpura	45	8865	50.0	443250	9850	97	955450
Tadatalav	5	42	50.0	2100	420	5	2100
Kantiajal	40	10560	50.0	528000	13200	78	1029600
Neja	15	2530	50.0	126500	8433	59	497567
Nada	116	33501	50.0	1675050	14440	253	3653342
Total	402	98513	50.0	4925650	12253	881	10746735

Source: Primary Survey, 2003.

Table 5.7 presents the data on the man-days spent on collection of mangroves in the reference year along with the value at the prevailing wage rates in the selected villages. The total value of collection of mangroves for the sample households comes to Rs. 49.26 lakhs. The total value for the collection of mangroves for the selected villages (calculated for the mangrove users) comes to Rs. 107.47 lakhs.

Value of Mangroves as Fuel Wood and for Construction

The next important direct use of mangroves after fodder is for fuel wood and construction. As shown by table 5.8 about 24.2 per cent households (one fourth households) in coastal region use mangroves for fuel wood and about 10 per cent households use it for construction.

Table 5.8 shows that mangroves are popular fuel wood in Asirawandh, Gunao and Lukky (in Kachchh-Jamnagar region), in Bhangadh and Mahadevpura (Gulf of Khambhat) and Kantiajal, but not in Tadatalav, or Neja. This is perhaps because the supply of mangroves is not even in all the region. Also, mangroves are not used for construction (of boats, house etc.) in Tadatalav and Neja-Nada, perhaps because the mangroves here are stunted and are not good enough for the use of construction.

Table 5.8

Advantages of Mangroves as Fuel Wood as Perceived by Users
(From House Listing)

Villages	% of HH Using Mangroves as Fuel Wood	% of HH Using Mangroves for House Construction	% of Users with Opinion			
			Good Fuel	Burn Even When Wet	Less Smoke	No Special Advantages
Asirawandh	86.7	0.0	76.9	30.8	69.2	15.4
Gunao	13.3	31.7	75.0	25.0	50.0	0.0
Lukky	1.0	3.6	0.0	0.0	0.0	0.0
Bhangadh	52.3	26.9	44.2	20.3	54.1	1.7
Mahadevpura	28.6	11.3	0.0	0.0	4.9	0.0
Tadatalav	0.0	0.0	0.0	0.0	0.0	0.0
Kantiajal	0.0	0.4	0.0	0.0	0.0	0.0
Neja	0.0	0.0	0.0	0.0	0.0	0.0
Nada	2.0	0.0	0.0	0.0	0.0	0.0

Source: Primary Survey, 2003.

Table 5.8 shows that people use mangroves as fuel wood because it is a good fuel—it burns even when it is wet and emits lee smoke while burning. It needs to be added, however, that there is no charcoal making from mangroves in Gujarat as it is seen in some other parts of the world.

Table 5.9 presents information about the use of mangroves as fuel wood by the selected households in the different seasons of the year. The table makes it clear that mangroves are used as fuel wood mainly in summer and winter. The use is the highest in Gunao in the sense that all the households in the village use mangroves as fuel wood. The use is also high in Bhangadh and Mahadevpura, Asirawandh and Kantiajal. The other fuel wood used are Ganda Baval (in all the seasons), followed by cow dung. A few households also use kerosene, LPG etc.

Table 5.9

Seasonal Consumption of Mangroves as Fuel in Selected Villages

Village	Sample HH	% of HH Using as Fuel in Summer				
		Mangrove	Ganda Baval	Cow Dung	LPG	Other
Asirawandh	14	57.1	100	71.4	0.0	0.0
Gunao	25	100	100	100	4	0.0
Lukky	12	16.7	100	0.0	0.0	0.0
Bhangadh	130	62.3	99.2	50	0	36.2
Mahadevpura	45	77.8	100	48.9	2.2	53.3
Tadatalav	5	0.0	100	100	0.0	0.0
Kantiajal	40	12.5	100	57.5	2.5	45
Neja	15	0.0	100	66.7	46.7	13.3
Nada	116	6	87.1	71.6	3.4	20.7
Total	402	41	96	60	3	29

Village	Sample HH	% of HH Using as Fuel in Monsoon				
		Mangrove	Ganda Baval	Cow Dung	LPG	Other
Asirawandh	14	0.0	100.0	71.4	0.0	0.0
Gunao	25	4	36.0	76.0	4	0.0
Lukky	12	0.0	91.7	0.0	0.0	0.0
Bhangadh	130	3.1	55.4	50.0	0.0	36.2
Mahadevpura	45	0.0	60.0	48.9	2.2	53.3
Tadatalav	5	0.0	100.0	100.0	0.0	0.0
Kantiajal	40	0.0	50.0	57.5	2.5	40.0
Neja	15	0.0	73.3	66.7	46.7	13.3
Nada	116	0.0	68.1	61.2	2.6	18.1
Total	402	1	62	56	3	27

Village	Sample HH	% of HH Using as Fuel in Winter				
		Mangrove	Ganda Baval	Cow Dung	LPG	Other
Asirawandh	14	35.7	100.0	71.4	0.0	0.0
Gunao	25	100.0	100.0	100.0	4.0	0.0
Lukky	12	16.7	91.7	91.7	0.0	0.0
Bhangadh	130	62.3	99.2	50.0	0.0	36.2
Mahadevpura	45	77.8	100.0	48.9	2.2	53.3
Tadatalav	5	0.0	100.0	100.0	0.0	0.0
Kantiajal	40	12.5	100.0	57.5	2.5	45.0
Neja	15	0.0	100.0	66.7	46.7	13.3
Nada	116	6.0	85.3	71.6	2.6	20.7
Total	402	40	95	63	3	29

Source: Primary Survey, 2003-04.

The annual consumption of fuel wood from mangroves in the selected villages has been calculated in Table 5.10. This has been calculated for the households who use mangroves as fuel wood. As the table shows, the annual use of mangroves for fuel wood per household varies from 0 in Neja to 626 kg. in Mahadevpura. It is 429 kg. in Bhangadh, 405 kg. in Gunao, 231 kg. in Asirawandh and 19 kg. in Kantiajal, while the average for all households it is 257.6 kg.

The monetary value of mangroves for fuel has been calculated by using the average price of fuel wood per kg prevailing in the villages. This price comes to Rs. 2.00 per kg. (Table 5.10). The total value of the consumption of this fuel wood for the selected households comes to Rs. 7.54 lakhs.

This value has been scaled up for the total area under mangroves in the state by calculating region-wise consumption of mangroves for fuel wood. As presented by Table 5.11, this comes to 4.49 crores.

Table 5.10

Annual Consumption and Value of Mangrove Used as Fuel Wood by Villagers

Villages (Name)	Sample HH (No.)	Total Use ('000 Kg.)	Average Use per HH (Kg.)	Total Consumption of the Village ('000 Kg.)	Market Value (Rs./ Kg.)	Total Value (Lakhs of Rs.)
Asirawandh	14	3.24	231	6.94	2	0.14
Gunao	25	10.12	405	24.29	2	0.49
Lukky	12	0.48	40	1.12	2	0.02
Bhangadh	130	55.73	429	189.48	2	3.79
Mahadevpura	45	28.16	626	133.29	2	2.67
Tadatalav	5	0.0	0.0	0.0	2	0.0
Kantiajal	40	0.74	19	4.40	2	0.09
Neja	15	0.0	0.0	0.0	2	0.0
Nada	116	5.10	44	17.63	2	0.35
Total	402	103.57	257.6	377.16	2	7.54

Source: Estimated from Primary Survey, 2003.

Table 5.11

Total Value of Mangroves for Fuel Wood for the State

Regions	Value of Mangroves as Fuel per ha.**	Total Area under Mangroves in the Region (ha.)*	Total Value of Fuel for the Year (Lakhs of Rs.)
Kachchh and Jamnagar	462	84900	392.2
Saurashtra	462	1700	7.9
Gulf of Khambhat	1578	3000	47.3
South Gujarat	88	1500	1.3
Total	2590	91100	448.8

Source: *: FSI (2001), *Survey of Forest Report*
　　　　**: Estimated from Primary Survey, 2003.

Value of Mangroves as Timber

Value of mangroves as timber has been calculated, again, on the same lines. As shown in Table 5.8, there is a practice of using mangroves as timber for construction work—of fishing boats, house etc. However, the use is limited because the mangroves in many areas are short and stunted. Consequently, though many households reported that they use mangroves for construction purposes (Table 5.8), the actual use was very low, negligible in most villages. In fact, only one region reported significant use of mangroves. The value of the mangrove as timber used by people in Gujarat coast comes to Rs. 17.10 lakhs.

The current value of mangroves used for medicine, for vegetation is also noted in the field survey. However, the use is very small and therefore not quantified.

Collection of Crabs, Mudskippers and Fish from Mangrove Waters

Though all the villages reported that they collect crabs, mudskippers etc from mangrove waters for self consumption or for sale, no concrete data were available about the quantity or the value of such collections. The base line study by GUIDE, however, includes a study on this for the Kachchh area. According to this study, per hectare output of crabs and mudskippers in mangrove waters is about 400 kg. per year. Assuming that the value remains the same in all

Table 5.12

Value of Mangroves as Timber

Regions	Value of Mangroves as Timber per ha. of Mangrove Cover (Rs.)	Total Area under Mangroves in the Region (ha.)*	Total Value for the Region per Year (Rs.)
Kachchh and Jamnagar	0	84900	0
Saurashtra	0	1700	0
Gulf of Khambhat	577	3000	1731000
South Gujarat	0	1500	0
Total	577	91100	1731000

Source: Estimated from Primary Survey, 2003.

mangrove areas, we have calculated the value for the state. At the rate of Rs. 20 per kg., the monetary value comes to Rs. 72.88 crores (Table 5.13).

Table 5.13

Annual Value of Collection of Crabs, Mudskippers etc. in Mangrove Waterways

Regions	Total Area under Mangroves in the Region (ha.)*	Value of Crabs, Mudskippers etc. per ha.	Total Value per Year (In Crores of Rs.)
Kachchh and Jamnagar	84900	8000	67.92
Saurashtra	1700	8000	1.36
Gulf of Khambhat	3000	8000	2.4
South Gujarat	1500	8000	1.2
Total	91100	8000	72.88

Source: Estimated from Primary Survey, 2003.

Total Direct Use Value of Mangroves in Gujarat

The total (current) annual value of the products of mangroves emanating from their direct use has been presented in Table 5.15. As the table shows, the total value comes to Rs. 160.3 crores. The most important source of direct value fodder, which contributes about 51.6 per cent of the total direct value. This is followed by mudskippers and crabs in mangrove waters (45.5 per cent), fuel wood (2.8 per cent) and construction (0.1 per cent). It needs to be noted that this direct value is occurring when the quality of mangroves is not very satisfactory in the state.

Table 5.14

Estimated Direct Use Value of Mangroves per ha. per year

Regions	Per ha. Value for the Year (in Rs.) for				
	Fodder	Fuel Wood	Construction	Crabs and Mudskipper	Total
Kachchh and Jamnagar	8689	462	0	8000	17151
Saurashtra	8689	462	0	8000	17151
Gulf of Khambhat	13114	1578	577	8000	23269
South Gujarat	23860	88	0	8000	31948

Source: Estimated from Primary Survey, 2003.

Table 5.15

Direct Use Value of Present Area Under Mangroves in Crores of Rs. (for 2003)

Regions	Value of Fodder	Value of Fuel	Value of Mudskippers and Crabs	Value of Construction	Total Direct Use Value	% Share
Kachchh and Jamnagar	73.77	3.92	67.92	0.00	145.62	90.8
Saurashtra	1.48	0.08	1.36	0.00	2.92	1.8
Gulf of Khambhat	3.93	0.47	2.40	0.17	6.98	4.4
South Gujarat	3.58	0.01	1.20	0.00	4.79	3.0
Total	82.76	4.49	72.88	0.17	160.30	100.0
% Share	51.6	2.8	45.5	0.1	100.00	

Source: Estimated from Primary Survey, 2003.

Direct Use Value of Mangroves for the Potential Areas

In addition to the current direct use value of mangroves, we have also calculated direct use value for the areas that have been identified as potential areas for mangroves in the state. This has been done using the per ha. values of mangroves for different direct uses, calculated on the basis of the field survey. These are presented in Table 5.14. Based on these rates, the total direct use value for the potential area has been calculated in Table 5.16.

Table 5.16

Direct Use Value for Potential Area Under Mangroves

(in Crores of Rs.)

Regions	Value of Fodder	Value of Fuel	Value of Mudskippers and Crabs	Value of Construction	Total Direct Use Value	% Share
Kachchh and Jamnagar	40.59	2.16	0.00	37.37	80.11	65.54
Saurashtra	1.77	0.09	0.00	1.63	3.50	2.86
Gulf of Khambhat	13.91	1.67	0.61	8.48	24.68	20.2
South Gujarat	10.41	0.04	0.00	3.49	13.94	11.4
Total	66.68	3.96	0.61	50.97	122.22	100.0
% Share	54.55	3.24	0.5	41.7	100.00	

Source: Estimated from Primary Survey, 2003.

The total direct use value of mangroves: The direct use value of mangroves for the current year for the potential area is Rs. 122.22 crores. If the current direct use value added to it, the total direct use value comes to Rs. 282.53 crores.

Indirect Use Value of Mangroves

The major indirect use values arising from the ecological functions of mangroves can be listed as follows:

- Its value in fisheries—onshore and offshore fisheries (fish, shrimp, prawns, crabs etc.)

- Its value for protection to agriculture from winds, cyclones, salinity etc.

- Its value for providing protection to life, property infrastructure etc. to coastal settlements.

- Its value for promoting biodiversity in coastal and marine ecosystems as well as improving quality by coastal water and groundwater.

It is clear that this value will increase multifold with the improvement in the quantity and quality of mangroves on the coast of Gujarat.

It has not been possible to compute these values, as some of the methods of valuation are beyond the purview of this study. We have therefore (1) compiled values wherever it is possible and (2) used the results of the available studies when conducting studies has not been possible.

Indirect Use Value of Mangroves for Onshore Fishery

The discussion in the earlier section has shown that fishery is an important occupation in these villages. In the Kachchh villages about 20 per cent households have fisheries as their main occupation, while 15 per cent households have taken it up as secondary occupation. In Kantiajal also about 15 per cent households have fishery as their main occupation, followed by Nada (13 per cent) and Neja (15 per cent). In Bhangadh and Mahadevpura fisheries is taken up mainly as secondary occupation by some households.

The development of fisheries depends on the availability of fish, which to a considerable extent depends on the status of mangroves. Mangroves supply food to marine communities via a detritus food chain starting from fallen mangroves leaves. They also provide a habitat for some commercially exploited marine organisms like prawn, crabs, fish etc. at critical phases of their life cycle by functioning as a feeding ground for juveniles (nurseries). The areas with thick mangroves therefore, have good fishing industry and the areas with degraded and sparse mangroves have poor fishing industry. As noted by Chhaya, "mangrove estuarine systems provide large nutrients to waters, with productivity measurements as high as 8.23 per sq. meter per day! Leaf and twig litter of mangroves of approximately a metric tones per hectare. Per year represents a major portion of detritus-based aquatic food chain in which much of the organic production is exported to surrounding waters, which in turn support highly productive near shore fisheries." (Chhaya, 1997).

The valuation of mangroves for promoting fisheries requires knowledge about: (1) the relationship between the quality of mangrove ecosystem and fish etc. (i.e. the biological dose response relationship), (2) the catch function, i.e. the function describing catch as a function of population sizes and human efforts, (3) the demand curve for fish etc., and (4) the supply curve and how it shifts under the influence of a change in stocks. However, all these data are not easily available. Such data are not available even with the department of fisheries. We have therefore, used a simple method here. We have assumed that the mangroves contribute about 50 per cent of the value of the fishery caught by fishermen. This assumption is based on empirical studies conducted in other parts of the world.

Table 5.17

Value of Fish-catch in the Selected Villages

Villages	Fishing HH	Fish-catch per HH per Year	Total Fish-catch of the Village for the Year	Average Value of Fish	Value of Fish-catch for the Year
(Name)	(No.)	(kg.)	(kg.)	(Rs./kg.)	(Rs.)
Asirawandh	7	300	2100	30	63000
Gunao	11	190	2090	30	62700
Lukky	2	775	1550	30	46500
Bhangadh	12	350	4200	30	126000
Mahadevpura	6	600	3600	30	108000
Tadatalav	4	200	800	30	24000
Kantiajal	36	270	9720	30	291600
Neja	5	700	3500	30	105000
Nada	52	795	41340	30	1240200
Total	135	500	68900	30	2067000

Source: Estimated from Primary Survey, 2003-04.

The total value of the fish-catch in the selected villages has been estimated at Rs. 20.67 lakhs. The indirect use value of mangroves in fishery can be put at Rs. 10.38 lakhs for the year 2003.

Indirect Use Value of Onshore Fisheries for the Present Cover of Mangroves and with Restoration of Mangroves in the Potential Area

Table 5.18-A presents the indirect use value of mangroves for fisheries for the present area under mangroves. This value, which has been calculated on the basis of the *Primary Survey*, comes to Rs. 8.69 crores.

The indirect use value of mangroves, in terms of fish production, has been calculated for the potential areas under mangroves at the present rate of fish-catch by the households engaged in fisheries. The total indirect value for the potential area under mangroves has been presented in Table 5.18-B, according to which the value is Rs. 5.99 crores. The total value, that includes present area under mangroves as well as the potential area, will come to Rs. 15.68 crores.

Table 5.18

Indirect Use Value of Mangroves: For On-site Fishery

A. Value of On-site Fish-catch in the Present Cover of Mangroves

Regions	Value of Fish-catch per ha (Rs.)*	Total Area Under Mangroves in the Region (ha.)	Total Value for the Region per Year (In Crores of Rs.)
Kachchh and Jamnagar	957	84900	8.1
Saurashtra	957	1700	0.2
Gulf of Khambhat	658	3000	0.2
South Gujarat	1458	1500	0.2
Total	4030	91100	8.7

B. Value of On-site Fish-catch in the Potential Area of Mangroves

Regions	Value of Fish-catch per ha (Rs.)*	Total Potential Area for Mangroves in the Region (ha.)	Total Value for the Region per year (In Crores of Rs.)
Kachchh and Jamnagar	957	46707	4.5
Saurashtra	957	2039	0.2
Gulf of Khambat	658	10606	0.7
South Gujarat	1458	4363	0.6
Total	4030	63715	6.0

Source: * Estimated from Primary Survey, 2003 (Fifty per cent of actual value was taken for calculation)

Value of Mangroves for Offshore Fisheries

Studies show that a mangrove forest of 1 ha. can support more than 1 tonne of fish production of off shore fishery per ha. per year (Schatz, 1991). With this assumption we have calculated the amount of off-site marine fish-catch that is likely to be supported by mangrove forest for the present area as well as the potential area. Table 5.19-A shows the total value of off-site fish-catch supported by present mangrove cover in the state, which comes to 221.4 crores per year and is about 12.5 per cent of the total marine fisheries production of the state for the year 2002-03. Similarly, the value of off-site fisheries for the potential area of mangroves in the state has been calculated in Table 5.19-B, which comes to about Rs. 154.8 crores. If the current indirect use value of off-site fishery is added, it comes to a total of Rs. 376.2 crores.

Table 5.19

Indirect Use Value of Mangroves: For Off-site Fishery

A. Value of Off-site Fishery for the Present Cover of Mangroves

Regions	Amount of Off-site Fish-catch per ha. (kg.)	Total Present Area Under Mangroves in the Region (ha.)	Total Amount for the Region per year (kg.)	Average Value of Fish (Rs./kg.)*	Total Value for the Region per year (In Crores of Rs.)
Kachchh and Jamnagar	1000	84900	84900000	24.3	206.3
Saurashtra	1000	1700	1700000	24.3	4.1
Gulf of Khambat	1000	3000	3000000	24.3	7.3
South Gujarat	1000	1500	1500000	24.3	3.6
Total	4000	91100	91100000	24.3	221.4

B. Value of Off-site Fishery for the Potential Area of Mangroves

Regions	Amount of Off-site Fish-catch per ha. (kg.)	Total Potential Area for Mangroves in the Region (ha.)	Total Amount for the Region per Year (kg.)	Average Value of Fish (Rs./kg.)*	Total Value for the Region per Year (In Crores of Rs.)
Kachchh and Jamnagar	1000	46707	46707000	24.3	113.5
Saurashtra	1000	2039	2039000	24.3	5.0
Gulf of Khambat	1000	10606	10606000	24.3	25.8
South Gujarat	1000	4363	4363000	24.3	10.6
Total	4000	63715	63715000	24.3	154.8

Source: * Estimated from the total marine fish production and its value in state for year 2002-2003 as quoted in *Socio-Economic Review Gujarat State: 2003-2004.*

Indirect Use Value of Mangroves for Protection

Indirect use value of mangroves for protection of life and property of people and for protection of agricultural lands from saline waters and saline winds has been calculated using the WTP approach. A question was asked to all households as to how much is the value of mangroves for the village for the protection it provides to life and property and for the protection it provides to its agricultural lands. The question about the protection to life etc. was replied by the households on the basis of their experiences in the past and their perception about the protection that mangroves would provide to the village. The village-wise values presented in the Table 5.20 is the average of the values given by the households. The question about protection to agricultural lands was replied by households in terms of their individual assessments for their respective farms. These were totalled up and scaled up to arrive at village level estimates.

Table 5.20

Indirect Use Value of Mangroves for Protection in Sample Villages

Villages	Average Value of Protection from Cyclone (in Lakhs of Rs.)	Protection Cost of Agricultural land (in Lakhs of Rs.)	Total Cost (in Lakhs of Rs.)
Asirawandh	1.5	0.0	1.5
Gunao	0.7	0.8	1.5
Lukky	1.3	0.4	1.7
Bhangadh	10.6	10.1	20.7
Mahadevpura	10.3	3.9	14.3
Tadatalav	6.3	3.5	9.8
Kantiajal	12.2	6.3	18.5
Neja	1.0	1.2	2.1
Nada	11.0	22.1	33.2
Total	54.9	48.4	103.2

Source: Estimated from Primary Survey, 2003.

Table 5.21

Indirect Use of Mangroves

A. Estimated Indirect Use Value per ha. of Mangroves Cover

Regions	Value of Mangrove Cover (Rs. Per Ha.) for			
	Protection from Cyclone	Protection Cost of Agricultural	Value of On-site Fish-catch	Value of Off-site Fish-catch
Kachchh and Jamnagar	3839	2039	957	24300
Saurashtra	3839	2039	957	24300
Gulf of Khambhat	3569	3275	658	24300
South Gujarat	12175	6307	1458	24300

B. Total Indirect Use Value of Mangroves: For the Present Cover of Mangroves

Regions	Total Value (In Crores of Rs.)				
	Protection from Cyclone	Protection of Agricultural Land	On-site Fish-catch	Off-site Fish-catch	Total
Kachchh and Jamnagar	32.59	17.31	8.12	206.31	264.33
Saurashtra	0.65	0.35	0.16	4.13	5.29
Gulf of Khambhat	1.07	0.98	0.20	7.29	9.54
South Gujarat	1.83	0.95	0.22	3.65	6.64
Total	36.14	19.59	8.70	221.37	285.80

C. Total Indirect Use Value of Mangroves: For the Potential Area of Mangroves

Regions	Total Value (In Crores of Rs.)				
	Protection from Cyclone	Protection of Agricultural Land	On-site Fish-catch	Off-site Fish-catch	Total
Kachchh and Jamnagar	17.93	9.52	4.47	113.50	145.42
Saurashtra	0.78	0.42	0.20	4.95	6.35
Gulf of Khambhat	3.79	3.47	0.70	25.77	33.73
South Gujarat	5.31	2.75	0.64	10.60	19.30
Total	27.81	16.16	6.00	154.83	204.80

Source: Estimated from Primary Survey, 2003.

The total indirect use value of mangroves will be Rs. 490.6 crores (Rs. 285.8 crores + Rs. 204.8 crores).

It needs to be underlined here that the perceived value of mangroves as a protective shield against strong winds, cyclones and tsunami waves etc. depends on the experience or the living memory of the population. For example, if the population has experienced a major crisis in the recent past, they are likely to put a very high value to the protection. The present study was conducted before the tsunami disaster, but if it was conducted after the tsunami disaster, the value given to mangroves as a protective wall against cyclones etc would have been much higher.

Value of Mangroves for their Ecological Functions

As seen earlier, mangroves perform several important ecological functions like promoting biodiversity in coastal and marine ecosystems as well as improving quality by coastal water and ground water. However, valuation of these functions requires a larger study, which is beyond the scope of this study. The present study therefore, uses some other relevant studies for the purpose of valuation of the ecological functions of mangroves.

Potential Direct Use Value of Mangroves

Mangroves in Gujarat are used for limited purpose only. However, there are many other uses, which can contribute significantly to the total use volume of mangroves. These uses in Gujarat are not tapped because: (1) there is not enough knowledge about the uses or the commercial promotion techniques of tapping the uses; (2) enterprising population of the state has not yet taken interest in mangroves and their uses, or because (3) mangroves in Gujarat are depleted and scattered, which makes it difficult to use mangroves for these additional uses.

In this section, value of mangroves will be calculated for all the potential uses of mangroves.

Wood Products

Timber: Some species of mangroves, including the *Avicennia* (i.e. *A. Officinalis, A. Marina, A. Aeba* etc.) which is predominant in Gujarat and

Rhizophora which is grown in some parts of state, can produce good quality timber, when the mangroves achieve good growth.

Mangroves are already used in the state for construction of fishing boats and houses/huts when the growth of mangroves is good. This potential of mangroves can be tapped to its maximum when mangroves acquire good growth. For example, in Singapore unsawn poles of *Rhizophora* (bakan piles) are the most common extraction product of mangroves. Similarly, in Malaysia also mangroves are used as timber in a big way. It has been estimated in the context of Vietnam that the value of mangroves per hectare as timber will be equivalent to Rs. 7560 per ha per year (Binh and Hoang, 1999). This implies that mangroves can generate Rs. 117.04 crores worth timber per year in Gujarat if they achieve good growth in the present and potential areas of mangroves.

Fuel Wood-Charcoal: Mangroves fuel wood can carry more value when converted into charcoal. *Avicennia* and *Rhizophora* wood (both growing in Gujarat) have a high calorific value meaning that they produce relatively more heat for the same weight as compared to other wood and can therefore, produce good charcoal. Mangrove based charcoal is produced in Indonesia, Malaysia and Thailand. Gujarat can also produce charcoal from mangroves once mangroves grow very well. The annual net return per hectare of charcoal obtained from mangroves forests is estimated to be equivalent to Rs. 13591 (Sathirathi, 2000). The total revenue that can be generated from mangroves based charcoal in the state will be about Rs. 210.4 crores.

Raw Material for Industries: Mangroves are also used as raw material for a number of wood based industries, like chip board, pulp wood (news papers and card board), synthetic materials (e.g. rayon) etc.

Non-wood Products

Food: Mangroves can be used as human food. The seedlings as well as seeds of *Avicennia* and *Bruguiera* species are consumed by people in several parts of the world.

Tannins and Dyes: Bark of mangroves are harvested as a source of tannin for tanning industry, as high tannin content is found in some species of mangroves, particularly *Rhizophora*. Mangroves can also be used to make black dye for dyeing cloth in some East African Countries. Many tannin factories in Bangladesh depend on mangrove extracts. This use of mangrove is also observed in Sundarbans, where about 10,000 tonnes of mangroves bark is harvested annually for the purpose.

Medicinal Use: Mangroves are useful as medicinal plants in many parts of the world. The following table presents the major medicinal use of mangroves.

Table 5.22
Some Medicinal Uses of Mangroves

Name of Plant	Plant Part	Medicinal Importance
Acanthus ilicifolius Linn.	Leaf extract	Treating rheumatism and neuralgia
Barringtonia racemosa (L.) Bl.	Roots	As de-obstruent
Caesalpinia bonduc (Linn.) Roxb.	Fruits	Cure for cough, asthma and diarrhoea
	Leaf paste	Cures on swollen testicles
	Leaves and bark	Posses anthelminthic properties
	Seeds	Cures jaundice
Calophyllum inophyllum L.	Seed oil	Used for curing rheumatism, skin disease and leprosy
Ceriops tagal (Perr.) C.B. Rob.	Decoction of shoots	Cure for malaria
Clerodendrum inerme (Linn.) Gaert.	Leaves	Cure for skin diseases
		Used as pain killer
Heritiera fomes Buch.-Ham.	Wood	Cure for piles
		used in boat(canoe) building, and house construction
Heritiera littoralis Dryn.	Decoction of seeds	Used for curing diarrhoea and dysentery
Hibiscus tiliaceus L.	Roots	Possess fabrifuge operative, emollient, psudorific, laxative and diuretic properties
	Leaves	For curing pimples
Pongamia pinnata (Linn.) Pierre	Bark and crushed seeds	Cure for malaria, skin diseases, rheumatic joints and leprous sores
Scaevola taccada (Gaertn.) Roxb.	Leaves	Used as fabrifuge,
		Cure for headache and cough
Terminalia catappa L.	Leaves	For treating rheumatic joints
	Leaf juice	For curing scabies and coetaneous diseases
	Leaf paste	Cure for skin diseases
Thespesia populnea (Linn.) Sol. ex Corroa	Leaves	Cure for stomach troubles
Xylocarpus granatum Koen.	Bark	Cure for dysentery
Xylocarpus moluccensis (Lamk.) Roem.	Bark	Cure for dysentery

Source: Swaminathan and Deshmukh, 1995

A study conducted in Indonesia (Ruitenbeek, 1992) established that medicinal plants from mangroves generate about US $ 15.00 medicinal value per hectare. It is a potential benefit of mangroves that can be captured if mangroves are well grown. Some other studies of mangroves based pharmaceutical values show that the value can go up to US $ 61.00 per hectare. That way in Gujarat the amount of revenue that can be generated from medicinal use of mangroves varies from Rs. 10.9 crores to Rs. 44.3 crores.

Honey and Bee Wax: Honey and bee wax are important products of mangroves in India as well as in some other countries like Bangladesh. In Bangladesh honey and pollen are used as medicines, high energy food and as a source of vitamins and minerals. Honey is collected using traditional methods and is sold to nearby communities (Basit, 1995). Honey and bee wax collection provides a seasonal source of income to many in Bangladesh and India.

Table 5.23

Sundarbans Honey and Beeswax Production and Revenue

Year	Honey (MT)	Honey Revenue (taka)	Beeswax (MT)	Beeswax Revenue (taka)
1990–91	211.27	536 400	52.8	211 200
1989–90	146.55	620 280	36.5	195 400
1988–89	99.45	84 560	24.9	39 840
1987–88	223.31	178 650	55.8	89 280
1986–87	229.11	183 930	57.5	92 040
1985–86	224.52	180 450	56.4	89 220
1984–85	255.80	102 800	64.2	51 390
1983–84	260.35	114 610	65.4	52 360
1982–83	232.65	93 460	58.12	46 730
1981–82	225.26	107 050	53.92	53 520
1980–81	310.93	120 450	75.03	60 030

Source: DFO, Sundarbans, Forest Department (As quoted in Basit, 1995).

Other Products: Mangroves have several other uses, such as, raw material in manufacturing of alcohol, vuinegar, gum, mosquitocides etc.; as a source of tea; for manufacturing feed for prawn etc. (Thangam, 1990; Premanathan, 1991; Kathiresan and Ramesh, 1992;

Moorthy, 1995; Kumaran, 1995; Palanisselvan, 1995; Kathiresan, 1995 etc.)

Wildlife Resources: Mangroves are a habitat for wild life like tigers (Sundarbans, West Bengal) as well as other animals and reptiles. These resources can produce meat, skin, fur etc. In addition they can also serve as attractive tourist sites.

Recreation and Ecotourism

Tourism is both a strong source of foreign exchange and a sustainable source of income if appropriate tourist sites can be developed and maintained in good condition. Mangrove areas can become important tourism and recreational sites.

Tobias and Mendelsohn (1991) used the travel costs method to estimate the value of Monteverde Cloud Forest Reserve in Costa Rica for ecotourism. Using the TCM, Tobias and Mendelsohn (1991) estimated a $35 per visitor value for recreation at a 10, 000 hectare Costa Rican tropical forest reserve. They included only Costa Rican visitors in their study. Costanza (1989) used two methods to calculate the value of coastal wetlands recreation in the the United States. Using the TCM, they estimated the value at $70.67 per visitor. Using the CVM (Contingency Valuation Method), they estimated a value of $47.11 per visitor. Following are some studies with estimated value of mangroves forests in terms of Ecotourism.

Table 5.24

Economic Value of Ecotourism in Mangroves Forests
as per Previous Studies

Study	Year	Area	Value (Rs. per ha. per annum)
Bennet and Reynolds	1993	Sarawak	19652
Costanza	1997	Global	22990
Leong	1999	Kuala Selangor, Malaysia	42874

In Gujarat the Marine National Park of Jamnagar is an ideal place for developing ecotourism. With proper planning much of the area under dense mangroves in Gujarat can be developed for ecotourism. In the case of Gujarat, two spots, namely, Kachchh and Jamnagar, can be developed as tourist spots in the near future, while in the long run, two more spots, at Bhavnagar and Surat can be developed. The last two sites have the potential of dense growth of mangroves in the long run. We have used the average value of above three studies i.e. Bennet and Reynolds (1993), Costanza *et al.* (1997) and Leong (1999) for estimating the probable value of ecotourism that can be developed in Gujarat in near future (Table 5.25-A). The value comes to about Rs. 41.6 crores per annum. In long run the value may go up to Rs. 115.14 crores (Table 5.25-B).

Table 5.25

Potential Value of Mangroves for Ecotourism

A. In Near Future

Districts having Potential for Development of Ecotourism (Name)	Dense Mangroves* (Ha.)	Value of Ecotourism per ha. per annum (Rs.)	Total Value of Ecotourism per annum (Crores of Rs.)
Kachchh	11800	28506	33.64
Jamnagar	2800	28506	7.98
Total	14600	28506	41.62

B. In Long Run

Districts having Potential for Development of Ecotourism (Name)	Present Cover of Dense Mangroves* (Ha.)	Dense Mangroves in Potential Area in Long Run** (Ha.)	Value of Ecotourism per ha. per Annum (Rs.)	Total Value of Ecotourism per Annum (Crores of Rs.)
Kachchh	11800	16365	28506	80.29
Jamnagar	2800	5575	28506	23.87
Bhavnagar	1000	1000	28506	5.70
Surat	800	1050	28506	5.27
Total	16400	23990	28506	115.14

Source: *: Singh, H.S. (2000).

**: It has been assumed that in long run there will be dense mangroves cover in at least 50 per cent of the potential area in respective regions.

In a majority of protected areas in India, tourism is low either due to the fact that these places are unknown to all but the most avid of wildlife lovers, or because of a complete absence of facilities for tourists. For well-known national parks economic value of ecotourism is quite high. (Table 5.26).

Table 5.26

Economic Value of Ecotourism in National Parks in India

Location	Methodology Applied	Value in Rs. per ha. per Year	Source
Keoladeo National Park, Bharatpur	Travel Cost Method	16197	Chopra (1998)
Keoladeo National Park	Contingent Valuation Method	20944	Murthy & Menkhuas (1994)
Boriveli National Park, Mumbai	Contingent Valuation Method	23300	Hadker *et al.* (1995)

Source: *TWG Economics and Valuation of Biodiversity*, Thematic Draft BSAP, 2002.

Summary Picture of Direct and Indirect Use Value

The summary picture of direct and indirect use value is presented in the Tables 5.27 and 5.28 show that the total present direct and indirect use value of mangroves is Rs. 773.13 crores per year.

Table 5.27

Direct Use Values of Mangroves

Goods and Services	Value per Annum (in Crores of Rs.)			
	From Present Area of Mangroves	From Potential Area of Mangroves	Total Direct Use Value	Value per ha. per Annum (in Rs.)
Direct Use Values				
Fodder	82.76	66.68	149.44	9652.83
Fuel Wood	4.49	3.96	8.45	545.92
Mudskippers and Crabs	72.88	0.61	73.49	4747.10
Total Value of Construction	0.17	50.97	51.15	3303.63
Indirect Use Values				
Value of Protection:				
Protection Property from Cyclone	36.14	27.81	63.95	4130.96
Protection of Agricultural Land	19.59	16.16	35.75	2309.26
Value of On-site Fish-catch	8.70	6.00	14.70	949.41
Value of Off-site Fish-catch	221.37	154.83	376.20	24300.00
Total Direct and Indirect Value	446.11	327.03	773.13	49939.12

Source: Estimated from Primary Survey.

Table 5.28

Potential Use Values of Mangroves Goods and Services

Goods and Services	Value per Annum (in Crores of Rs.)	Value per ha. per Annum (in Rs.)
Wood Products:		
Timber	117.04	7559.99
Fuel Wood-Charcoal	210.40	13590.41
Raw material for Industries:	-	-
Non-Wood Products.		
Food	-	-
Tannins and Dyes:	-	-
Medicinal Use	44.30	2861.48
Honey and Bee Wax	-	-
Other Products	-	-
Wildlife Resources	-	-
Ecotourism:	115.14	7437.26
Total	486.88	31449.15

Source: Estimated from Secondary Sources.

In addition, the potential direct use value of mangroves is Rs. 486.88 crores per year. In other words, the total use value of mangroves per year will be Rs. 1260.01 crores per year. This is an under estimation as it does not include the potential indirect use value of mangroves.

6

Estimating Non-use Value of Mangroves

Non-use value of mangroves arises from the mere existence of mangroves. It basically refers to existence value, bequest value, intrinsic value, stewardship value and option value. As discussed earlier, existence value refers to willingness to pay for preservation of a resource, now or in the future. Option value refers to the value people attach to the environment just to hold or preserve or have the option by using it in the future, they may not know the value of the resources at present. Bequest value refers to the willingness to pay by people to preserve mangroves for the sake of their future generations, while stewardship value refers to the value people attach to mangroves as stewards of mangroves.

Methodologically speaking, these values can be computed using contingency valuation, under which people are asked their Willingness To Pay (WTP) for the preservation of mangroves, irrespective of its present or future use for them. Thus this value can be construed to be having some altruistic motive, whether pure or impure, whether paternalistic or non-paternalistic. Altruism is basically a satisfaction to people that mangroves exist.

The term 'existence value', first introduced by Weisbrod (1964), indicates that people value existence of things like unspoilt wilderness, and this utility can be valued in terms of willingness to pay. This 'passive use value' or the utility people attach to the existence of unspoilt wilderness is similar in kind to, and directly comparable with, the value from active use of wilderness area for tourism, logging and so on. The non-use value highlights the distinction between the 'use value' by visiting an area for recreation and the values expressing willingness to pay for preservation of an area they will never visit. Those who support contingent valuation approach use the term passive use value, and those who are skeptical about contingent valuation, use the term 'existence value' or 'non-use value'.

According to Mitchell and Carlson (1989), respondents to contingent valuation studies generally offer about four times the once and for all payment for preservation as they often when asked for an annual payment. This suggests a discount rate of 25 per cent or, more plausibly, a general inability to undertake present value calculations. Thus the existence value is the value the individual derives from the knowledge that the site exists, even if he/she never plans to visit it. In other words, existence value is the value derived from the sheer contemplation of the existence of ecosystem, apart from any direct or indirect uses of goods and services they provide. This can include a pure biodiversity component, the appreciation for the variation or richness in the ecosystems. This is based on the contemplation of the ecosystem as a whole versus appreciation of each of its members individually.

Quiggin I (1998) has subdivided the existence value into sub-categories: (1) psychic or vicarious consumption; (2) option value; (3) altruism; (4) bequest value; (5) intrinsic value, and (6) stewardship value. As discussed earlier, these values refer to the different, frequently overlapping, dimensions of the non-use value of mangroves.

The non-use value of mangroves can be captured by estimating WTP of non-users. As Mitchell and Carlson (1989) have argued, "It should be kept in mind that the only valid measure of the non-use value of a natural resource is the willingness to pay (WTP) amounts of non-users." This is because, (1) the non-use value occurs largely to non-users of the resource (though users are not excluded) and (2) it is difficult for users to specify as to which portion of WTP for the use value and which portion is of non-use value.

Since the valuation of non-use value of mangroves required contingency valuation from non users, to spread across the state or even country, the valuation could not be undertaken by us.

In fact, the monetary valuation of non-use value of mangroves requires a separate study altogether. Since the existence value is the willingness to pay even when direct on-site interactions between the user and the mangroves does not occur, this interaction is non-consumptive, indirect and off-site. In short, existence value is reflected in the fees and voluntary contributions by members of environmental groups for preservation of mangroves irrespective of access. That is, the existence value of mangroves will refer to the willingness to pay for preservation of mangroves even when there is no physical access to the mangroves.

In this study we have estimated three types of non-use values namely,

- Non-use value of mangroves for carbon sequestration;

- Non-use value of mangroves for coastline stabilisation, and

- Non-use value of mangroves for preservation value.

Non-use Value and Indirect Use Value

It needs to be underlined, however, that it is not always easy to distinguish between the non-use value and indirect use value of a natural resource. Conceptually speaking, the non-use value of a natural resource can be distinguished clearly from its indirect use value, based on the definitions of the two concepts, but it is difficult to do so in reality. This is primarily because the non-use value tends to become an indirect use value, once the importance of the natural resource is appreciated. For example, the Bequest Value or the Option Value of mangroves can be perceived as its non-use value, once it is realised that preservation of mangroves have an important role to play in the ecology. That is, non-use value frequently tends to turn in to indirect use value. In this chapter we have discussed three types of non-use values as mentioned above. However, these can be covered under 'indirect use value' of mangroves.

Carbon Sequestration

The relationship between climate change (one of today's leading environmental concerns) and the conservation and development of the world's forest has become a major issue today. Tropical forests, including mangroves, have an important role in regulating carbon dioxide in the global atmosphere through the processes of respiration and photosynthesis, whereby plants absorb carbon oxide/dioxide and store it in their biomass. A major ecological function of mangroves is to serve as carbon sink.

The general approach in estimating the potential of a forest in sequestrating carbon involves calculating the total biomass per hectare (biomass density), and then applying appropriate conversion factors to get the carbon equivalents. Several studies have been conducted on above-ground biomass and productivity of mangrove forests in terms of carbon sequestration (e.g., Golley *et al.*, 1962; Lugo and Snedaker, 1974; Briggs, 1977; Christensen, 1978; Suzuki and Tagawa, 1983; Tamai *et al.*, 1986; Day *et al.*, 1987; Lee, 1990; Kusmana *et al.*, 1992; Saintilan, 1997a & b), but few on the below-ground biomass (Golley *et al.*, 1962; Lugo and Snedaker, 1974; Briggs, 1977; Komiyama *et al.*, 1987; Saintilan, 1997a & b).

Estimation of the potential of a forest in sequestrating carbon involves calculating the total biomass per hectare (biomass density), and then applying appropriate conversion factors to get the carbon equivalents. Ong, Gong and Clough (1995), estimated the biomass and productivity in a 20 year old stand of *Rhizophora apiculata* Mangrove Forest as 7.14 tC per ha. per year.

Table 6.1

Proportion of Biomass and Productivity in a 20-year Old Stand of Rhizophora Apiculata Mangrove Forest

		% Biomass	*Productivity (tC/ha/Year)*
Canopy	Leaves	2.6	0.08
	Branches	8.0	0.44
Trunk		74.0	5.56
Stilt Roots		10.0	0.64
Roots		5.1	0.42
Total Biomass		**100.0**	**7.14**

Source: Based on Ong, Gong and Clough (1995).

Mangrove forests usually create thick, organically rich sediments as their substrata. Most of the substrata in the tropics except under deltaic environments consist of mangrove peat which mainly derives from mangrove roots (Scholl, 1964a & b; Woodroffe, 1981; Fujimoto and Miyagi, 1993; Fujimoto *et al.*, 1995b; Fujimoto *et al.*, 1996). This shows that mangrove forests have great below-ground productivity and play a significant role in carbon sequestration not only above ground but also below ground.

The rate of carbon sequestered in mangrove mud is estimated to be around 1.5 t C ha-1 yr-1 (Ong, 1993). The upper layers of mangrove sediments have a high carbon content (a conservative estimate is 10 per cent). Each hectare of mangrove sediment would then contain some 700 tonnes of carbon per metre depth (Ong, 2002)

In estimating a monetary value of the carbon sequestered by the forest, an international price per unit amount of carbon reduced will have to be applied. For this study, the adopted price is $150 per tonne of carbon (based on the tax rate in Norway). The indirect use value of mangroves in terms of carbon fixation thus adopted in this

Table 6.2

*Value of Mangroves for Carbon Sequestration in
Present and Potential Area*

(In Crores of Rs.)

Regions	Value of Carbon Sequestration in Present Cover of Mangroves per Annum from			Value of Carbon Sequestration in Potential Area of Mangroves per Annum in		
	Biomass of Mangroves*	Mangroves Sediment**	Total (Biomass + Sediment)	Biomass of Mangroves*	Mangroves Sediment**	Total (Biomass + Sediment)
Kachchh and Jamnagar	421.45	88.54	509.99	231.86	48.71	280.57
Saurashtra	8.44	1.77	10.21	10.12	2.13	12.25
Gulf of Khambat	14.89	3.13	18.02	52.65	11.06	63.71
South Gujarat	7.45	1.56	9.01	21.66	4.55	26.21
Total	452.23	95.01	547.23	316.29	66.45	382.73

Source: *: Based on Ong, Gong and Clough (1995).
    **: Based on Ong (1993).

study includes both carbon fixations in the mangrove biomass as well as in the mangroves sediment. The total value for carbon sequestration in mangroves forest in Gujarat has been calculated in Table 6.2.

The total value of Carbon Sequestration for the present and potential area of mangroves in Gujarat comes to about 929.97 crores per annum.

Shoreline Stabilisation/Erosion Control

When land is a traded commodity, the value of the shoreline stabilisation functions of mangroves might involve estimating the land area lost due to erosion, and valuing that loss at the current land price (i.e., the value of land lost). The value of property lost as a result of shoreline erosion might also be estimated. Where land is not traded, an appropriate technique involves valuing production (e.g., of agriculture) from that land and estimating the lost net output if erosion persists (i.e., the change in productivity approach). Using the later approach, Ruitenbeek (1991) estimates the benefit of erosion control of mangroves at Rp. 1.9 million per household per year for Bintuni Bay, Indonesia. In some cases mangroves may actually lead to land accretion. Any such positive additions of land should be added to that saved through shoreline stabilisation and erosion control. The avoided cost approach might also be used to measure the benefits of soil erosion control. This might involve, for example, estimating the construction and operating cost of a system of dams, weirs, artificial reefs, or other 'engineered' solutions to avert erosion.

Decision makers often undervalue the shoreline protection service that natural landscapes provide, and often don't take it into account in the decision making process. Part of the reason is that there are no quantitative measures of this service. In this respect following facts are the eye openers:

- Japan has identified approximately 46 per cent of its shoreline as requiring stabilisation measures, and during 1970–98 the total investment on shoreline protection works was 4.5 trillion Yen (more than US$ 40 billion) (Japan Ministry of Construction 1998).

- Since 1965, the United States has spent US$ 3.5 billion on 1,305 beach replenishment projects. For example, the beach replenishment of Miami Beach in the late 1970s alone cost US$ 64 million (Williams *et al.* 1995).

- It is estimated that the state of New Jersey would require US$1.6 billion over 10 years to replenish and maintain its 90 miles of developed open ocean shoreline (Trembanis *et al.* 1998: 246–251).

- According to the Harbour Department of the Ministry of Communication and Transport, Thailand, several areas along the coastline that have no mangrove cover, experience severe erosion and require breakwater construction. The unit cost of constructing this type of dam is 35,000 Baht per metre of coastline. Based on above information, the replacement cost to protect the shoreline when there is destruction of a strip of one Rai (0.16 ha) of mangrove with a 75 metre-width along the coastline is approximately 746,666.7 Baht (US$ 29,866.67). The annualised value (through the project life of 20 years) is therefore, 37,333.3 Baht (US$ 1,493.33) per Rai. According to the Harbour Department, approximately 30 per cent of the coastal areas experience severe erosion requiring the construction of breakwaters. Thus Sathirathi (1997) adopted 12,444 Baht (US$ 478.63) per Rai (Rs. 137606 per ha) as a proxy for the value of mangrove in terms of coastline protection. However, it should be noted that there is a tendency for the replacement cost in this case to be overestimated since building this type of dam to protect the coastline does not use up as much land area as it would if it was left under mangrove cover. These opportunity costs of land are not taken into account.

Based on the study of Sathirathi (1997) we have calculated the value (cost) of mangroves for shore line stabilisation and erosion control in Gujarat as below.

The total value of erosion control for the total area under mangroves comes to Rs. 2130.34 crores.

Table 6.3

Non-use Value of Mangroves: For Erosion Control

Regions	Value of Coastline Protection in Crores of Rs. for		
	Present Cover of Mangroves	Potential Area for Mangroves	Total Area of Mangroves
Kachchh and Jamnagar	1168.27	642.72	1810.99
Saurashtra	23.39	28.06	51.45
Gulf of Khambat	41.28	145.94	187.23
South Gujarat	20.64	60.04	80.68
Total	1253.59	876.76	2130.35

Source: Estimated from Secondary Sources.

Mangroves for the Future

It needs to be noted that people of the coastal Gujarat do realise the importance of mangroves, though they may not be aware of all the benefits of mangroves and therefore, they may not appreciate the full value of mangroves. The coastal villages consider mangroves extremely important for their life and livelihood. Mangroves are important in their culture and traditions also. In many villages we were told that "Cheriyas (mangroves) are our life" and "tears come to our eyes if they are destroyed." Under the REMAG project all the villages have formed local societies for regenerating mangroves, "to make the sea coast rich with mangroves ultimately." Our investigation revealed that all households, whether users or non-users of mangroves directly, are interested in their restoration and regeneration. Without a single exception all households want thick forest of Cheriyas around their villages.

The households want to expand the areas under mangroves by (1) plantation and regeneration work—as being done under the REMAG project, (2) protection of mangroves from over grazing and by (3) controlling its over use by awareness generation among people and by producing alternatives (like grass /fodder) of mangroves (Table 6.4).

Because of the work done by the NGOs in these villages, they are aware about the need for protection of mangroves from indiscriminate grazing and cutting. As shown in Table 6.4, majority of the village households believe that projects for restoration and regeneration of

Table 6.4

Villagers' Opinion on How Mangroves Can be Increased?

Villages	% Total	% of HH with Opinion			
		Plantation and Regeneration Projects	Protection & Maintenance	Awareness Generation	Can't Say
Asirawandh	100	71	21.4	7.1	0
Gunao	100.0	72.0	4.0	20.0	4
Lukky	100.0	91.7	8.3	0.0	0
Bhangadh	100.0	46.2	42.3	6.2	5.4
Mahadevpura	100.0	73.3	26.7	0.0	0.0
Tadatalav	100.0	60.0	20.0	0.0	20.0
Kantiajal	100.0	85.0	7.5	7.5	0.0
Neja	100.0	60.0	26.7	6.7	6.7
Nada	100.0	50.9	37.9	0.0	11.2

Source: Primary Survey, 2003.

Figure 6.1

How Can We Save Mangroves? Villagers Opinion

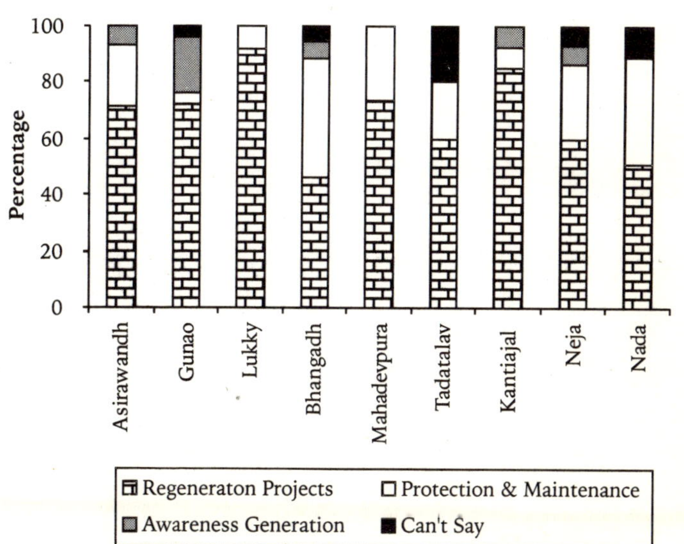

Source: Primary Survey, 2003.

mangrove should be taken up for expanding mangroves forests on the coast. Though the households have not been able to say how much more area needs to be covered under mangroves, the general reply is 'large areas', 'more than the present area' or 'as much area as possible'. An important implication of these replies is that the coastal villages are keenly interested in the expansion of mangroves areas over the coast. They also want that mangroves should be protected from overuse and destruction and should be utilised in a sustainable manner. Though it is accepted that the users of mangroves can not perceive adequately the non-use value of mangroves, the above responses, to a considerable extent, reflects the non-use value as perceived by the villagers.

While concluding this chapter, we would like to observe that the non-use value of mangroves is quite high. Though we could not measure it specifically from non-users of Gujarat or India, it is very high, which makes mangroves a highly valuable natural resource.

7

Concluding Observations and Inferences

One major observation emerging from the study is that mangroves are extremely valuable forest ecosystem. Its various economic uses and ecological functions make it very important to the ecology and economy of any country. Unfortunately, its value is far from appreciated thanks to the non-market nature of most of its value, which fails to enter into national or state income accounts. As a result, mangrove has become a neglected ecosystem in most countries. This study, however, has shown that the monetary value of mangroves, if calculated carefully, is highly significant, which implies that mangroves should receive a high priority in development policy and planning.

Mangroves have been generally neglected in India, resulting in a significant decline in the area under mangroves over the past decades. The present mangroves cover is therefore less than 4500 sq. km. area over the long sea coast of more than 8000 km. Gujarat also has experienced a drastic decline in the area under mangroves during the past 50 years or so. Though Gujarat has the longest sea coast (of 1650 km.) in the country, mangroves cover only about 911 sq. km. Area (FSI, 2001), of which about 80 per cent is located in Kachchh district. In other words, most of the Gujarat coast has experienced severe depletion and degradation of mangroves.

The value of mangroves can be basically seen as its replacement value and as its total economic value, which includes its use value (direct and indirect use value) and non-use value. The former refer to the cost of plantation and protection of mangroves till it reaches maturity, while the latter includes the range of use and non-use values. These values can also be seen as preservation and development values. One can compute these values by developing suitable concepts and methods, which is what has been attempted here.

It is not easy to compute the monetary value of the different economic and ecological functions of mangroves due to several reasons. Firstly, the value of mangroves occurs not only to local communities, but also to the state, to the country and to the world. Rich mangroves on the Gujarat coast can contribute significantly to the country and the world by enriching biodiversity and promoting carbon sequestration. There is a need to compute these values in monetary terms to estimate the total value of mangroves. Secondly, the value of mangroves increases at an exponential rate with their growth. That is, the value of mangroves increases multifold for all its functions when patchy, stunted and under grown mangroves acquire maturity. It is important to incorporate this exponential growth in value while compiling total value of mangroves. Thirdly, mangroves sustain and promote the livelihood of millions of people on the (Gujarat) coast by providing fodder for their animals, fuel wood and timber for their consumption, fish for their livelihood and protection to their life by protecting them from saline and strong winds. The monetary value of this livelihood protection cannot be captured adequately by the money value of these uses. The livelihood dimension or the distribution aspect of the benefits goes beyond the money value of mangroves. And lastly, it is not easy to estimate the protection provided to life and property of coastal population by a mangrove bio-shield against cyclones and tsunami kind of disasters. Though we have tried to incorporate some of these points in the valuation, it has not been possible to include all the issues in the valuation. In that sense the valuation done here is under valuation.

Replacement Value of Mangroves: In this study, we have first of all calculated the replacement cost of mangroves for the loss experienced during 1998-2001 as well as for the long term loss experienced during the last fifty years or so. The results show that the loss of mangroves experienced by the state during 1998-2001 can be met by spending Rs. 23.30 crores on its regeneration. The full restoration of mangroves in the state (by planting mangroves also on the potential area under mangroves) will cost about Rs. 118 crores. That is, by spending Rs. 118 crores, the state will be able to restore mangroves in the state. This is not a very high amount, particularly in the context of the benefits received from mangroves.

Use and Non-use Value of Mangroves

The benefits of mangroves include the use and non-use values of mangroves at different levels as well as the potential value of mangroves. Our study has shown the following:

- The direct use value of mangroves for the present area under mangroves in the state is Rs. 160.30 crores per year (2003). The direct use value for the potential area of mangroves is Rs. 122.22 crores per year. This implies that the state can gain Rs. 282.53 crores per year (at 2003 prices) from mangroves as its direct use value. It needs to be noted that this value is low because it has been calculated for the present state of mangroves. With the improvement in the quality of mangroves, this value will rise at an exponential rate.

- The indirect use value of the current status of mangroves is Rs. 285.80 crores per year, while the same value for the potential area is Rs. 204.80 crores per year. The total indirect use value of mangroves thus will be Rs. 490.60 crores per year. This value is also low and with the improvement in the quality of mangroves, it will rise at an exponential rate.

- In short, the total use value of mangroves is about Rs. 773.13 crores per year for the state (at 2003 prices). That is, the state can earn Rs. 773.13 crores per year as use value of mangroves by restoring mangroves in the state. This will rise as the quality of mangroves improves.

- The use of mangroves, however, is limited because the other uses are not yet tapped by people of Gujarat due to several reasons like poor growth of mangroves, lack of information/technology etc. Mangroves have many more uses, which primarily refer to the various wood based products, non-wood products and ecotourism etc.

- It has been estimated that the wood products of mangroves (for timber and fuel wood—charcoal) alone can generate an income of about Rs. 327.44 crores per year. This excludes other industrial uses of timber based product like pulp wood, chips etc, for which we do not have adequate database.

- Mangroves can be used in the production of several non-wood products like tannin and dyes, medicine products, honey and wax and many other products. It is not easy to calculate the potential values of these products in the absence of the required data. We have some data on the medicinal values from a study in Indonesia, according to which the medicinal value (potential) of mangroves can go up to Rs. 44.30 crores per year.

- Gujarat has two mangrove sites, which can be developed, as ecotourism sites in the near future. These sites are at Kachchh and Jamnagar. The ecotourism value, the potential increase that can be generated from these sites, has been estimated to be Rs. 41.6 crores per year. (Based on other studies). Two more sites can be developed within 5-7 years at Bhavnagar and Surat, which can also generate similar amounts for the state. In long run the value can go up to 115.14 crores per year.

- In short, the total potential use value of mangroves can be around Rs.486.88 crores per year. Again, this is gross underestimation as we do not have the required data for a large number of products of mangroves.

- The non-use value of mangroves is extremely important due to the immense contribution of mangroves to the ecology of the state, country as well as the world. Two sets of values, namely, carbon sequestration and shoreline stabilisation, have been estimated using the studies conducted in other countries. It has been estimated that the present value of carbon sequestration is about Rs. 547.23 crores per year while the same for the potential mangroves area will be Rs. 382.74 crores per year. In other words, both the value, put together will come to about Rs. 929.97 crores per year.

- The non-use value of mangroves as shoreline stabiliser has been estimated at Rs. 1253.59 crores for the present area and Rs. 876.76 crores for the potential area, bringing the total to Rs. 2130.35 crores per year. The total non-use value thus comes to Rs. 3060.31 crores per year.

In short, the value of the benefits—use and non-use values—generated by mangroves at present in Gujarat is about Rs 2246.93 crores per year. Thus one can say that by spending Rs. 118 crores, we can generate more than Rs. 4000 crores (from present as well as potential area for mangroves) from mangroves resources of Gujarat. This is an underestimation because of the factors mentioned above as well as because we have not been able to value several of the mangrove products and several ecological functions adequately. For example, the different wood and non-wood product values of mangroves, as discussed earlier, are not included because of the lack of the data. Also, the entire contribution of mangroves in terms of protecting and enhancing biodiversity has been excluded—except for the inclusion of onshore and offshore fisheries. Mangroves are one of the best promoters and supporters of biodiversity. For the lack of the required data, however, we have not been able to estimate this value.

Figure 7.1

Total Value of Mangroves in Gujarat for Present Area and Potential Area

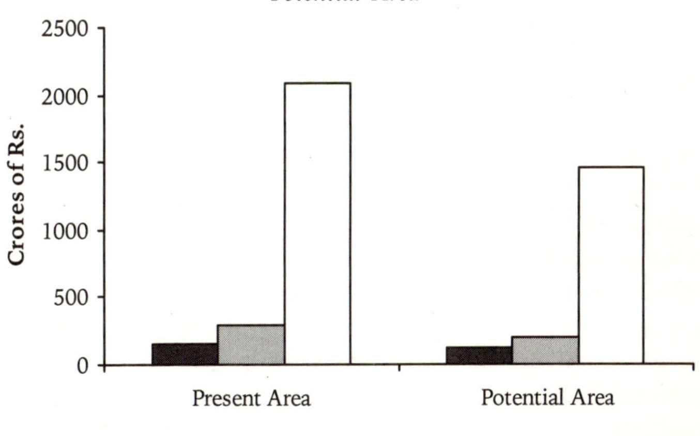

Source: Based on Primary Survey.

This value will or can increase multifold due to several factors: (1) the quality of mangroves will improve with its growth, as a result of which its productivity will rise at an exponential rate, (2) with the improvement and expansion of mangroves, it will be possible to use it for a number of new products and (3) the increased cover and improved quality of mangroves will enhance its ecological functions. At the rate of 10 per cent increase in the value per year, one can expect the total value of Rs 13,535.00 crores produced by mangroves within five years!

Similarly, the mangroves planted on the potential areas will generate additional Rs. 586.52 crores of value per year. We do not calculate the incremental value here, as this will be new area for mangroves. Within the coming five years this mangroves can generate the value of about Rs. 7930 crores. Adding both the values, one can say that mangroves can produce Rs. 12,465.00 crores worth value for the economy within the coming five years. Once again, this will be an under estimation as we have left out several areas of valuation as mentioned above.

Limitations of Valuation

It is extremely important to underline two points at this stage. Firstly, an underlying assumption of the valuation of mangroves is that there is substitutability between the ecological functions of mangroves and growth in the different sectors of the economy. This assumption, however, is not realistic because some of the ecological functions of mangroves that are critical for the ecological health of the state and the country or the world cannot be performed by any other means! For example, the value of mangroves for carbon sequestration cannot be adequately estimated in monetary terms, as this function can be performed only by mangroves/forests. Secondly, the contribution of mangroves in terms of protecting the livelihood of hundreds of people engaged in fishery and animal husbandry in the coastal regions of Gujarat cannot be adequately captured by the valuation, as the distributional aspect of the benefits cannot be measured adequately by the valuation.

Need For Further Research

This study is a methodological study in the sense that it has developed concepts and methods of estimating the value of a coastal ecosystem. It has shown that it is feasible to calculate the monetary value of the different economic and ecological contribution made by a coastal ecosystem.

The study has also revealed certain major data gaps for the purpose of valuation. Though it has filled in some of the data gaps arbitrarily by using the results of the available studies, many of them conducted outside India, it is clear that India has to develop this database for the country. India has more than 8000 km. of the coast and it is absolutely necessary that we develop our own database on the coastal ecosystems.

We therefore, suggest that further research is conducted in the following areas:

- A study of valuation of mangroves for the entire coast of India. This will give broad estimates of the contribution of mangroves to the state and the national economies.

- Such a study needs to focus also on valuing the biodiversity sustained and promoted by the mangroves ecosystems. This will contribute significantly to the literature on valuation of natural resources in India. Some methods on this are being developed in the literature and these need to be strengthened through conducting a well designed study.

- There is also a need to go one step further and conduct a comprehensive study on non-use value of mangroves in India. The non-use value will include carbon sequestration, shoreline stabilisation, biodiversity including value as a critical habitat for threatened species, water filter services etc. We strongly recommend this study, as we consider it extremely relevant and useful for understanding the contribution of mangroves to the ecology and economy of India.

Inferences for Coastal Development Policy

Poor valuation or no valuation of mangroves in our national income has resulted in our highly inadequate understanding not only of the ecology but also of the economy of the coastal regions. Decisions are taken about the development of this region without proper understanding either of the economy or of the ecology of the region. As a result, the delicate balance between the coastal/marine ecology and the economy seems to have been lost. On the one hand, one observes depletion and degradation of coastal/marine resources, while on the other hand one also observes shortage of drinking water, severe natural disasters, decline in agriculture and inadequate growth of employment/livelihood opportunities in coastal areas. A major policy implication of the study is that any policy or project that affects mangroves or coastal resources adversely, needs to be weighed systematically against the environmental costs.

Sustainable development essentially means integrating environmental resources or natural capital in to the development process so that the natural capital grows along with the economy. Mangroves is a natural capital, which has a potential of contributing significantly to the development of the coastal region in multiple ways. There is a need to explore this possibility of development, which promotes both the economy and the ecology.

Selected References

Aodgeson, G. and J.A.Dixon (1988). "Logging versus Fisheries and Tourism in Palawan", East-West Environmental Policy Institute Occasional Paper No.7, East-West Center, Honolulu.

Alam, A.M.M. Nurul (1992). *Revised Draft Working Plan of the Sundarbans Forest Division*, by Zillur Rahman, DFO, W.P. Division, Dhaka.

Ali, S.S. (1994). "Sundarban: Its Resources and Ecosystem", Paper presented at the National Seminar on *Integrated Management of Ganges Flood Plains and Sundarban Ecosystem*, 16–18 July.

Anon (Act Now for Ocean Natives (1988). *Mangroves in India: Status Report*, Government of India, Ministry of Environment and Forest, New Delhi, p.150.

––––––. (1992). *Mangrove Natural Products Valuation: Progress Report*, Philippine Institute for Development Studies, Makati, Matro Manila, December.

––––––. (1998). *Annual Data Book*. Can Gio Districts.

Aylward, B. (1991). "3. Biological Diversity (GK91-03)", *The Economic Value of Ecosystems*, Gatekeeper Series London Environmental Economic Center, p.10.

Azariah, J., V. Sevam, S. Gunasekaran (1992). "Impact of Past Management Practices on the Present Status of the Muthupet Mangroves Ecosystem", in Jaccarini, V. and E. Martens (eds.), *The Ecology of Mangroves and Related Ecosystems*, Vol. 247(1-3), pp.253-259.

Bagla, P. and S. Menon (eds.) (1989). Ravaged Forests and Soiled Seas: Ecological Issues in the Tropics with Special Reference to Andaman and Nicobar Island, Kalpavrish, New Delhi.

Bahuguna, Anjali, H.B. Chauhan and Shailesh Naik (1997). *Coastal Vegetation of Gujarat. Workshop on Integrated Coastal Zone Management*, Space Application Centre (ISRO), Ahmedabad, pp.208-221.

Bandyopadhya, A.K. (1991). "Role of Mangroves in Island Ecosystems with Particular Reference to the Bay Islands", Sanjay V. Deshmukh and R. Mahalingam (eds.), *A Global Network of Mangrove Genetic Resource Centres*, Project Formulation Workshop, pp.35-42.

Banerjee, J. (1957). "Regeneration and Exploitation of Mangroves", in *Mangrove Symposium*, Calcutta. pp.124-128.

Banerjee, V. and P.K.R. Choudhury (1987). "Preliminary Studies on Artificial Regeneration of Mangrove Forests in the Sunderbans, West Bengal", *Indian Forester*, Vol. 112(12), pp.203-222.

Banik, Swagata (2002). *Base Line Survey: Restoration of Mangroves in Gujarat*, Gujarat Ecology Commission, Vadodara.

Bann, Camille (1990). "The Economic Valuation of Mangroves: A Manual for Researchers", *Special Papers, International Development Research Centre*, Ottawa, Canada.

Bartelmus, Peter (1997). "The Value of Nature: Valuation and Evaluation in Environmental Accounting", *Working Paper Series No.15*, Department for Economic and Social Information and Policy Analysis, United Nations, New York.

Basha, S.C. (1991). "Distribution of Mangroves in Kerala", Indian Forester, Vol. 117(6), pp.439-448.

Basit, M.A. (1995). "Non-wood Forest Products from the Mangrove Forests of Bangladesh", in P.B. Durst and A. Bishop (eds.), *Beyond Timber: Social, Economic and Cultural Dimensions of Non-wood Forest Products in Asia and the Pacific*, RAP Publication 1995/13, FAO, Regional Office for Asia and Pacific, Bangkok.

Bateman, I. and Wills, K. (eds.) (1996). *Valuing Environmental Preference: Theory and Practice of the Contingent Evaluation Method in the US, EC and Developing Countries*, Oxford University Press, Oxford, UK.

Bennet, I.L. and C.J. Reynolds (1993). "The Value of a Mangrove Area in Sarawak", *Biodiversity and Conservation*, Vol. 2, pp.359-375.

Bhosale, L.J. and N.G. Mulik (1991). "Endangered Mangrove Areas of Maharashtra", in A.D. Agate, S.D. Bonde, K.P.N. Kumarna (eds.), *Proc. Symposium on Significance of Mangroves*, Maharastra Association for the Cultivation of Science Research Institute. pp.39-46.

Blasco, F. (1975). "Les Mangroves de L' Inde (The Mangroves of India)". Institute Francais De Pondichery, Trevaux be la section scientifique et Techhique. Tome XIV. Fascicule 1, pp.1-175.

————(1977). "Outline of Ecology, Botany and Forestry of the Mangals of the Indian Subcontinent", in V.J. Chapman (ed.), *Ecosystems of the World*, I-Wet Coastal Ecosystems. pp.241-262.

Brandon, Carter and Kisten Hommann (1995). *The Cost of Inaction: Valuing the Economy-Wide Cost of Environmental Degradation in India*, Asia Environment Division, World Bank.

Briggs, S.V. (1977). "Estimates of Biomass in a Temperate Mangrove Community", *Australian Journal of Ecology*, Vol. 2, pp.369-373.

Chakraborti, K. (1996). "Fish and Fish Resources in the Mangrove Swams of Sunderbans, West Bengal—An in-depth Study", *Indian Forester*, Vol. 112(6), pp.538-542.

Chakraborty, S.K. and A. Choudhury (1992). "Ecological Studies on the Zonation of Brachyuran Crabs in a Virgin Mangrove Island of Sundarbans, India", *J. Mar. Biol. Assoc. India*, Vol. 34(12), pp.189–194.

Chavan, S.A. (1985). "Status of Mangrove Ecosystem in Gulf of Kachchh", in *Symp. Endangered Marine Animals and Marine Parks, Marine Biological Society*, Paper No. 42, Cochin, India.

Chhaya, N.D. (1997). "Minding Our Marine Wealth: An Appraisal of Gujarat's Coastal Resources", *Environment and Development Series, Centre for Environment Education*, Ahmedabad.

Chmura, Gail L. (n.d.). *Carbon Sequestration in Mangrove and Salt Marsh Soils: Climatic Controls and Climate Feedback*, Abstract of Paper, Department of Geography, McGill University, Montreal, Canada.

Chopra, Kanchan (1993). "The Value of Non-timber Forest Products: An Estimation for Tropical Deciduous Forests in India", *Economic Botany*, Vol. 47(3), pp.251-57.

————. (2000). "Economic Valuation of Biodiversity: The Case of Keoladeo National Park", *Working Paper*, Institute of Economic Growth, Delhi.

Christensen, B. (1982). "Management and Utilization of Mangroves in Asia and the Pacific", *FAO Environment Paper No.3*, FAO, Rome.

Chopra, Kanchan and Gopal Kadekodi (1999). "Valuation of Forest Resources: An Application of the Contingent Valuation Method", *Working Paper Series*, Institute of Economic Growth, Delhi.

Coralie, Thornton, Mike Shanahan and Juliette Williams (2003). *From Wetlands to Wastelands: Impacts of Shrimp Farming*, SWS Magazine, Environmental Justice Foundation, London, UK. pp.48-53.

Constanza R., C. Farber and J. Maxwell (1989). "The Valuation and Management of Wetland Ecosystems", *Ecological Economics*, Vol. 1, pp.335-361.

Costanza, R., R. d'Arge, *et al.* (1997). "The Value of the World's Ecosystem Services and Natural Capital." *Nature*, Vol. 387 (May), pp.253-260.

CV-CIRRD (1993). "Mangrove Production and Management", *Central Visayas Technology Guide*. Cebu City, The Philippines, Central Visayas Consortium for Integrated Regional Research and Development (CV-CIRRD).

De Souza, N. and J. D'Souza, (1999). "Mangrove Sediment Enzymes: Indicator of Pollution", *Indian Journal of Environment Protection*, Vol.19(2), pp.132-136.

Deshmukh, S. and T.A. Rao (2003). "Mangrove Ecosystems of India: Status, Management and Policy", in Parikh J. and H. Datye (eds.), *Sustainable Management of Wetlands: Biodiversity and Beyond*. Sage Publication, New Delhi.

Deshmukh, S.V. (1991). "Description of the Site Selected for the Mangrove Genetic Resource Centre", in S.V. Deshmukh and R. Mahalingam (eds.), *A Global Network of Mangrove Genetic Resource Centres Project Formulation Workshop*, Madras, India, pp.99-106.

——. (1991). "Mangroves of India: Status Report", in S.V. Deshmukh and R. Mahalingam (eds.), *A Global Network of Mangrove Genetic Resource Centres Project Formulation Workshop*, Madras, India, pp.15-25.

Dick M. Melana, (2000). "Mangroves and their Importance to Food Security", *Over Seas: The Online Magazine for Sustainable Seas*, July, Vol.3(7).

FAO (1982), Management and utilization of mangroves in Asia and the pacific. Food and Agricultural Organization of the United Nations, Rome.

——. (1988). *Worldwide Compendium of Mangrove-associated Aquatic Species of Economic Importance*, FAO Fish. No. 814 (FIRI/C814), 236 pp.

——. (1994a). "Mangrove Forest Management Guidelines", *FAO Forestry Paper* No. 117, 319 pp. (issued also in Spanish).

Field, C.D. (1995). "Journey Amongst Mangroves", *International Society for Mangrove Ecosystems*, Okinawa, Japan.

——. (ed.) (1996). "Restoration of Mangrove Ecosystems", *International Society for Mangrove Ecosystems*, Okinawa, Japan.

FSI (2001). *State of Forest Report 2001*. Forest Survey of India, Ministry of Environment and Forest, Government of India, Dehradun, India.

Fujimoto, K., T. Miyagi, T. Kikuchi and T. Kawana (1996). Mangrove Habitat Formation and Response to Holocene Sea-level Changes on Kosrae Island, Micronesia. *Mangroves and Salt Marshes*, Vol. 1, pp.47-57.

Gammage, S. (1994). *Estimating the Total Economic Value of a Mangrove Ecosystem in El Salvador*. Report to the U.K. Overseas Development Administration (ODA), London.

GEC (2001). *State Environmental Action Programme, Sub-component Wetlands, Phase I Report*.

Gong, W.K. and J.E. Ong (1990). "Plant Biomass and Nutrient Flux in a Managed Mangrove Forest in Malaysia". *Estuarine, Coastal and Shelf Science*, Vol. 31, pp.519-530.

Gopal, B. and K. Krishnamurthy (1993). "Wetlands of South Asia", in Whigham, D.F., D. Dykyjova and S. Hejny *(eds.)*, *Wetlands of the World*, Vol. I: Handbook of Vegetation Science, Inventory, Ecology and Management, pp.345-414.

Goswami, Subhrangsu (2006). "Diminishing Global Mangrove Resources: Management and Conservation Issues in India", *CFDA Working Paper Series*, Working Paper No 12, CFDA, Ahmedabad.

Government of India (1997). "Mangroves of India: State-of-the-art Report (1987-1996)", *Environmental Information System Centre*, Centre of Advanced Study in Marine Biology, Annamalai University, Parangipettai, Tamil Nadu.

——. (1998). *Mangrove Atlas of India*, SAC, ISRO, Ahmedabad.

Government of Gujarat (2000). "Coastal Zone Information System: Gujarat", SAC (ISRO), Ahmedabad, RESECO, Gandhinagar and Forests & Environment Department.

Hamilton, L.S. and S.C. Snedakar (1989). *Hand Book of Mangrove Area Management*, p.123

Haque, Md Emdadul (2004). "The Sundarbans and Coastal Fisheries", *The Daily Star Web Edition*, May 28, Vol. 5(2).

Hirway, Indira & Subhrangsu Goswami, (2005). "Valuation of Mangroves in Gujarat", *Working Paper Series*, Working Paper No. 6, Centre for Development Alternatives, Ahmedabad

Hussain, K.Z. (1986). "Presidential Address", *Eleventh Annual Bangladesh Science Conference*, University of Rajshahi.

Islam, M.A. (1992). *Some Relevant Information about Sundarbans*, Sundarbans Forest Division, Government of the People's Republic of Bangladesh, Bangladesh.

IUCN (1999). "Economic Valuation of the Mangrove Ecosystem along the Karachi Coastal Areas", *The Economic Value of the Environment: Cases from South Asia*, IUCN.

Jagtap, T.G., V.S. Chavan and A.G. Untawale (1993). "Mangrove Ecosystems *of India: A Need for* Protection", *AMBIO*, Vol. 22(4), June, pp.225-254.

Jani, S.P. (2003). Status of Mangroves in the Gulf of Kutch along Jamnagar Coast, Paper presented in the *Workshop on Mangroves at NIO*, Goa.

Jayaseelan, M.J.P., V. Sundararaj, and M. Devaraj (1991). "Significance of Mangroves in Fisheries", in A.D. Agate, S.D. Bonde, K.P.N. Kumaran (eds.), *Proc. of the Symposium on Significance of Mangrove*, Pune, India, Maharashtra Assoc. Cultiv. Sci. Res. Inst., pp.14-23.

Jayasundaramma, B., R. Ramamurthi, E. Narasimhulu, and D.V.L. Prasad (1987). "Mangroves of South Coastal Andhra Pradesh: State of the Art Report and Conservation Strategies", in N.B. Nair (ed.), *Proc. of the National Seminar on Estuarine Management, Trivandrum*. pp.160-162.

Kapetsky, J.M. (1986). "Conversion of Mangroves for Pond Aquaculture: Some Short-term and Long-term Remedies", Paper presented at the *Workshop on the Conversion of Mangrove Areas to Aquaculture*, Iloilo City, The Philippines, 24-26 April, (*mimeo*).

Kathiresan, K. (1991). "Uses of Mangroves", *Yojana*, Vol. 31, pp.27-29.

——. (1993). "Dangerous Pest on Nursery Seedlings of *Rhizophora*", *The Indian Forester*, Vol. 119, p.1026.

——. (2000). *Mangrove Atlas in India*.

Kathiresan, K. and B.L. Bingham (2001). "Biology of Mangroves and Mangrove Ecosystems", *Advances in Marine Biology*, Vol. 40, pp.81-251.

Kathiresan, K. and M. Xavier Ramesh (1991). "Establishment of Seedlings of a Mangrove", *The Indian Forester*, Vol. 118, pp.687.

Kathiresan, K. and T. Subramonia Thangam (1987). "Biotoxicity of Excoecaria Agallocha L. Latex on Marine Organisms", *Curr. Sci.*, Vol. 56(7), pp.314-315.

Kapetsky, J.M. (1985). "Mangroves, Fisheries and Aquaculture", *FAO Fish. Report*, Vol. 338 (suppl.), pp.17-36.

Khalil, Samina (1999). "Economic Valuation of the Mangrove Ecosystem along the Karachi Coastal Areas", published in Joy E.Hecht (ed.), *The Economic Value of the Environment: Cases from South Asia*, IUCN.

Khan, A.S. (1994). *Bangladesh: Non-wood Forest Products in Asia*, Bangkok, FAO, Regional Office for Asia and Pacific, pp.1–8.

Kiyoshi Fujimoto (2000). "Below-ground Carbon Sequestration of Mangrove Forests in the Asia-Pacific Region", Paper presented at *International Workshop Asia-Pacific Cooperation Research for Conservation of Mangroves*, March 26-30, Okinawa, Japan.

Kothari, M.J. and K.M Rao (1993). *Environmental Impact on Mangroves of Gujarat-1BC 8*, pp.51-57.

Krishnamurthy, K. (1984). "The Conversion of Mangrove Lands and Water to Other Uses in India", Workshop on *Human Population, Mangrove Resources, Human Induced Stress and Human Health*. Bogo, Indonesia.

Krishnamurthy, K., A. Chaudhury and A.G. Untawale (1987). *Status Report—Mangroves in India*, Ministry of Environment and Forest, Govt. of India, New Delhi, p.150.

Krishnamurthy, K., K. Kathiresan, L. Kannan, N. Godhantaraman and W. Damodara Naidu (1995). Plankton of parangipettai (porto Novo), India Fasicle No.1. Microzooplankton of parangipettai (porto Novo), India Fasicale No. 1, Microzooplankton with special reference to Tintinnids, Annamalai University, pp.81.

Kulkarni, D.H. (1957). "Utilization of Mangrove Forests in Saurashtra and Kachchh", in *Mangrove Symposium*, Calcutta. pp.30-32.

Kumar, Ashwini (1996). *Environmental Issues in Integrated Coastal Zone Management for Jambusar, Bharuch, Gujarat*, School of Planning, CEPT, Ahmedabad.

Kusmana, C., S. Sabiham, K. Abe and H. Watanabe (1992). "An Estimation of above Ground Tree Biomass of a Mangrove Forest in East Sumatra, Indonesia", *TROPICS*, Vol. 4, pp.243-257.

Lal, P.N. (1990). "Ecological Economic Analysis of Mangrove Conservation: A Case Study from Fiji", UNDP/UNESCO Regional Mangrove Project RAS/86/120, *Mangrove Ecosystem Occasional Paper* No. 6, p.64.

Lee, S.Y. (1990). "Primary Productivity and Particulate Organic Matter Flow in an Estuarine Mangrove Wetland in Hong Kong", *Marine Biology*, Vol. 106, pp.453-463.

Leong, L.F. (1999). *Economic Valuation of the Mangrove Forest in Kuala Selangor, Malaysia*. M.Sc. Thesis.

Lugo, A.E. and S.C. Snedaker (1974). "The Ecology of Mangroves", *Annual Review of Ecology and Systematics*, Vol. 5, pp.39-64.

Macintosh, Donald J. and Elizabeth C. Ashton (2002). *A Review of Mangrove Biodiversity Conservation and Management*, Final Report, Centre for Tropical Ecosystems Research.

Mall, L.P., V.P. Sing, A. Garge and S.M. Pathak (1985). "Mangrove Forest of Andamans and Some Aspects of its Ecology", *Proc. Nat.Symp. Biol. Util. Cons. Mangroves*, Kohlapur. pp.438-443.

Mandal, R.N. and G.S. Sahu (2001). "Importance of Mangrove Vegetation in Cyclone-prone Coastal Region", *Journal of Environmental Research*, Vol. 11(2), pp.117-120.

Martosubroto, P. and N. Naamin, (1977). "Relationship between Tidal Forests (Mangroves) and Commercial Shrimp Production in Indonesia", *Marine Resources of Indonesia*, Vol. 18, pp.81-6.

Mathauda, G.S. (1957). "The Mangrove Forests of India", in *Mangrove Symposium*, Calcutta, pp. 66-87.

McCarthy, J.J., O.F. Canziani, N.A. Leary, D.J. Dokken and K.S. White (2001). "Climate Change 2001: Impacts, Adaptation and Vulnerability", Contribution of Working Group II to the *Third Assessment Report of the Intergovernmental Panel on Climate Change*, GRID, Arendal.

Moorty, M.N. (1999). "Environmental Values and National Income Accounting", *Discussion Paper Series* No. 9/99, Institute of Economic Growth, Delhi.

———. (2001). "Environmentally Sustainable Income and Shadow Prices of Natural Resources", *Working Paper Series* No. E/211/2001, Institute of Economic Growth, Delhi.

Moorthy, P. and K. Kathiresan (1995). "Influence of Nitrogen Salts on Growth and Physiological Responses of *Rhizophora apiculata* Blume in Non-aerated Water Culture", *Pakistan Journal of Marine Science*, Vol. 42 (in press).

Nambiar, Prithi. *The Report on The Marine National Park and Sanctuary, Jamnagar, Gujarat*, Centre for Environment Education, Ahmedabad

Naskar, Kumudranjan and Rathindranath Mandal (1999). *Ecology and Biodiversity of Indian Mangroves*, Daya Publishing House Delhi.

Nayak, Shailesh and Anjali Bahuguna, (2001). "Application of Remote Sensing Data to Monitor Mangroves and Other Coastal Vegetation of India", *Indian Journal of Marine Sciences*, Vol. 30(4), December, pp.195-213.

Ong, Jin Eong (1982). "Mangroves and Aquaculture in Malaysia", *Ambio*, Vol. 11, pp.252-257.

———. (1993). "Mangroves—A Carbon Source and Sink", *Chemosphere*, Vol. 27, pp.1097-1107.

———. (1995). "The Ecology of Mangrove Conservation and Management", *Hydrobiologia*, Vol. 295, pp.343-351.

———. (2002). "The Hidden Costs of Mangrove Services: Use of Mangroves for Shrimp Aquaculture", Background note prepared for *The International Geosphere-Biosphere Programme* (IGBP), Centre for Marine and Coastal Studies, University Sains, Malaysia.

Ong, Jin Eong and Gong Wooi Khoon, (1999). "Carbon Fixation in Mangrove Ecosystem and Carbon Credits", Workshop proceedings of *Traditional Knowledge and Emerging Technology*, Centre for Marine and Coastal Studies, University Sains Malaysia, Penang, Malaysia.

Ong, Jin Eong , W.K. Gong and B. Clough (1995). "Structure and Productivity of a 20 year-old Stand of *Rhizophora apiculata* Mangrove Forest", *J. Biogeography*, Vol. 22, pp.417-424.

Ong, Jin Eong, W.K. Gong and H.C. Chan (2001). "Governments of Developing Countries Grossly Undervalue their Mangroves"?, in UNEP-2001, *Proceedings of the International Symposium on Protection and Management of Coastal Marine Ecosystems*, pp.179-184, Bangkok, Thailand, 12-13 Dec. 2000, EAS/RCU UNEP Bangkok, Thailand, p.279.

Padmavathi, M. (1991). *Proceedings of National Symposium on Remote Sensing on Enviroment*, Anna University, Madras.

Palaniselvam, V. (1995). "Studies on the Cyanobacterium *Phormidium tenue* (Menegh) Gomant for its Utility in Aquaculture, Shrimp feed and as Biofertilizer for Mangroves", M.Phil. dissertation, Annamalai University, India, pp.45.

Paretta, J.C. (ed) (1993). *Marine Protected Area Needs in the South Asian Seas Region*, A Marine Conservation and Development Report. IUCN, Gland, Switzerland, pp.77.

Premanathan , M. (1991). "Studies on Antiviral Activity of Marine Plants", Ph.D. thesis. Annamalai University, India, pp.102.

Primavera, J.H. (1998). "Mangroves and Nurseries: Shrimp Population in Mangrove and Non-mangrove Habitat", *Coastal and Shelf Science*, Vol. 46, pp.457-464, Marine Science Institute, Estuarine.

Purushan, K.S. (1991). "Prospects of Fish Production from Mangrove Ecosystems", *Fish. Chimes*, Vol. 11(3), pp.24-26.

Rajagopalan, M.S., C.P. Gopinathan, V.K. Balchandran, and A.Kanagam (1986). "Prodctivity of Different Mangrove Ecosystems", *Proc.Symp. on Coastal Aquaculture*, Cochin 1980, Part 4: Culture of Other Organisms, Environmental Studies, Training, Extension and Legal Aspects. No.6, pp.1084-1087.

Ramdial, B.S. (1991). "Role and Importance of Mangrove Forests", in S.V. Deshmukh, R. Mahalingam (Eds.), *A Global Network of Mangrove Genetic Resource Centres Project Formulation Workshop*, Madras, India, pp.75-84.

Rao, A.N. (1991). "Evaluation, Utilization and Conservation of Mangroves:, in S.V. Deshmukh, R. Mahalingam (eds.), *A Global Network of Mangrove Genetic Resource Centres Project Formulation Workshop*, Madras, India, pp.75-84.

Ravikumar, S. (1995). "Nitrogen Fixing Azotobacters from the Mangrove Habitat and their Utility as Biofertilizers", Ph.D. thesis, Annamalai University, India. pp.202.

RSAM (Remote Sensing Application Misson) (1992). Coastal Environment. Space Application Centre (ISRO), Ahmedabad.

Ruddle, Kenneth and Walther Manshard (1981). "Renewable Natural Resources and the Environment", United Nations University, Tycooly International Publishing Limited, Dublin.

Ruitenbeek, H.J. (1992). "Mangrove Assessment: An Economic Analysis of Management Options with a Focus on Biutuni Bay, Irian Jaya", Environmental Management Development Indonesia (EMDI), *Environment Report*, Dalhousie University, Halifax, Canada, No. 8, p. 90.

Ruitenbeek, J. (1991). *Mangrove Management: An Economic Analysis of Management Options with a Focus on Bintuni Bay, Irain Jaya*, Ministry of State for Population and Environment, Jarkarta.

Saenger, P., E.J. Hegerl and J.D.S. Davie (1983). "Global Status of Mangrove Ecosystems", *IUCN Community Ecology Paper* No. 3, p.88 and *The Environmentalist*, Vol. 3 (3), pp.1-88.

Sahu, N.C., B. Nayak and D.P. Maharana (2000). "Economic Valuation of Environment: Components and Complexities", in N. Rajalaxmi (ed.), *Environment and Economic Development*, Manak Publications Pvt. Ltd., New Delhi.

Saintilan, N. (1997b). "Above- and Below-ground Biomass of Mangroves in a Sub-tropical Estuary", *Marine and Freshwater Research*, Vol. 48, pp.601-604.

Sathirathai, Suthawan (2000). "Economic Valuation of Mangroves and the Roles of Local Communities in the Conservation of Natural Resources: Case Study of Surat Thani, South of Thailand", *Research Reports*, International Development Research Centre, Ottawa, Canada.

Schatz, R.E. (1991). *Economic Rent Study for the Philippine,*Fisheries Sector Program, Asian Development Bank, Manila, pp.42.

Scholl, D.W. (1964a). "Recent Sedimentary Record in Mangrove Swamps and Rise in Sea Level over the Southwestern Coast of Florida: Part 1", *Marine Geology*, Vol. 1, pp.344-366.

———. (1964b). "Recent Sedimentary Record in Mangrove Swamps and Rise in Sea Level over the Southwestern Coast of Florida: Part 2", *Marine Geology*, Vol. 2, pp.343-364.

Schuyt, Kirsten and Luke Brander (2004). *The Economic Values of the World's Wetlands*, Prepared with support from the Swiss Agency for the Environment, Forests and Landscape (SAEFL) Gland/Amsterdam.

Scott, D.A., (1989). *A Directory of Asian Wetlands*, IUCN.

Sen Gupta, R. and Geetanjali Deshmukhe (2000). *Coastal and Maritime Environments of Gujarat: Ecology and Economics*, Gujarat Ecological Society, Vadodara.

Sen Gupta, R., Geetanjali Deshmukh, G.A. Thivakaran,K.Ramamoorthy, Rach Chandra, S. Serebiah and S. Bandopadhyay (1999). "Ecology of the Gulf of Kutch and Threat to Biotic Resources", Paper presented in the workshop on *ICMAM Plan for Gulf of Kachchh*, SAC, Ahmedabad.

Sharma, Devinder (2005). *Tsunami, Mangroves and Market Economy*, Web publication: *www.indiatogether.com*

Singh, A.P. (2002). "An Unique Terrestrial Ecosystem of White Mangrove (AVICINNIA OFFICINALIS LINN) in Kachch Dezert of Gujarat State", *The Indian Forester*, Vol. 128, No. 1, pp.95-98.

Singh, H.S. (1994). *Status Report on Mangroves in Gujarat State*, Gujarat Forest Department, Gandhinagar

———. (1999). *Mangroves in Gujarat*. Ganghinagar: Gujarat Ecological Education and Research (GEER) Foundation.

———. (2000). *Mangroves in Gujarat: Current Status and Strategy for Conservation*, Gujarat Ecological Education and Research (GEER) Foundation, Gandhinagar.

———. (2002). "Marine Protected Area in India: Status of Coastal Wetlands and their Conservation", Gujarat Ecological Education and Research (GEER) Foundation, Gandhinagar.

Singh, V.P., L.P. Mall, A. Garge and S.M. Pathak (1985). "Some Ecological Aspects of Mangrove Forest of Andaman Islands", *J. Bombay Nat. Hist. Soc.*, Vol. 83(3), pp.525–537.

Snedaker, S.C. (1978). "Mangroves: Their Value and Perpetuation", *Nature and Resources*, Vol. 15(3), pp.6 13, Unesco, Paris, France.

———. (1982). "A Perspective on Asian Mangroves", in C.H. Soysa, W.L. Colllei, C.L. Sien (eds.), *Man, Land and Sea: Coastal Resource Use and Development in Asia*, University of Singapore Press, Singapore, pp.65-74.

————. (1984). "Mangroves: A Summary of Knowledge with Emphasis on Pakistan", in B.V. Haq, J.D. Milliman (eds.), *Marine Geology and Oceanography of Arabian Sea and Coastal Pakistan*, Van Nostrand Reinhold Company, New York, pp.255-262.

————. (1984). "The Mangroves of Asia and Oceania: Status and Research Planning", in E. Soepadmo, A.N. Rao, D.J. McIntosh (eds.), *Proceedings of the Asian Mangrove Symposium*. Percetakan Ardyas Sdn Bhd., Kuala Lumpur, 5-15, Proc. Asian Mangrove Symp., August 25-29, 1980, Kuala Lumpur, Malaysia.

————. (1986). "Traditional uses of South American Mangrove Resources and the Socio-Economic Effect of Ecosystem Changes", in P. Kunstadter, E.C.F. Bird, S. Sabhasri (eds.), *Proceedings of the Workshop on Man in the Mangroves,* United Nations University, Tokyo, Japan, pp.104-112.

————. (1987). "Mangrove Forests of South America: Some Traditional Uses", *Bakawan*, Vol. 1, pp.10-12.

————. (1988). "Aquaculture and Mangrove Ecosystem Productivity in Arid and Semi-arid Coastal Environments", in M.F. Thompson, N.M. Tirmizi (eds.), *Marine Science of the Arabian Sea, Proceedings of an International Conference*, AIBS, Washington DC, pp.281-289.

————. (1989). "Mangroves: Productivity, Pollution and Planning", in M.R. Chim Figueiredo, N. Labbish Chao, W. Kirby-Smith (eds.), *Proc. International Symposium on Utilization of Coastal Ecosystems: Planning, Pollution and Productivity*, Vol. 2, da Fundacao Universidade do Rio Grande, Rio Grande, Brasil and Duke University Marine Laboratory, Beaufort, N.C, Beaufort, NC, pp.59-75.

Snedaker, S.C. and A.E. Lugo (1973). "The Role of Mangrove Ecosystems in the Maintenance of Environmental Quality and High Productivity of Desirable Fisheries", *Final Report on Contract No 14-16-008-606*, Bureau of Sport Fisheries and Wildlife, Atlanta, Georgia. 404 pages.

Snedaker, S.C. and M.S. Brown (1981). "Mangrove Misconceptions and Regulatory Guidelines", in R.C. Carey, P.S. Markovits, J.B. Kirkwood (eds.) *Proc. U.S. Fish and Wildlife Workshop on Coastal Ecosystems of the Southeastern United States.* (FWS/OBS-80/59) U.S. Fish and Wildlife Service, Office of Biological Services, Washington DC, pp.61-70.

————. (1981). *Water Quality and Mangrove Ecosystem Dynamics (EPA-600/4-81-002)*, US EPA, Office of Pesticides and Toxic Substances, Gulf Breeze, Florida, 79 pages.

————. (1982). "Primary Productivity of Mangroves", in C.C. Black and A. Mitsui (eds.), *CRC Handbook of Biosolar Resources*, "Vol. I: Fundamental Principles", CRC Press, Inc., Boca Raton, Florida, pp.477-485.

Snedaker, S.C. and P.D. Biber (1995). "Restoration of Mangrove Areas: A Case Study, Florida, USA", in C.D. Field (ed.), *Manual for Restoration of Mangrove Areas* (in review), International Society for Mangrove Ecosystems, Okinawa, Japan.

Snedaker, S.C., S.J. Baquer, P.J. Behr, S.I. Ahmed, (1995). "Biomass Distribution in Avicennia Marina Plants in the Indus River Delta", Pakistan, in M.F. Thompson, N.M. Tirmizi (eds.), *The Arabian Sea: Living Marine Resources and the Environment.* Vanguard Books (Pvt) Ltd., Lahore, Pakistan, pp.389-394.

Space Applications Centre (1997). *Proceedings of the Workshop on Integrated Coastal Zone Management*, ISRO, Ahmedabad and Department of Forest and Environment, Government of Gujarat.

————. (1998). "An Assessment of the Gujarat Cyclone Damage through IRS 1C/1D Data", *Scientific Report*, ISRO, Ahmedabad.

Spalding, M.D., F. Blasco and C. Field (1997). *World Mangrove Atlas*, The International Society for Mangrove Ecosystems, Okinawa, Japan, p.178.

Spaninks, Frank and Pieter Van Beukering (1997). "Economic Valuation of Mangrove Ecosystem: Potential and Limitations", *CREED Working Paper No. 14*, International Institute for Environment and Development (IIED), London and Institute of Environmental Studies (IVM), Amsterdam.

Stanford, R.L. (1973). "A Preliminary Literature Review on Mangroves: Soils, Phytogeography, Reproduction and Fauna", in S.C. Snedaker (ed.), *Ecological Studies on a Subtropical Terrestrial Biome*, UF/FP&L Report No. 1, University of Florida, Gainesville, Florida, 88.

Subramonia Thangam, T. (1990). "Studies on Marine Plants for Mosquito Control. Ph.D. Thesis, Annamalai University, India. pp.68.

Suzuki, E. and H. Tagawa (1983). "Biomass of a Mangrove Forest and a Sedge Marsh on Ishigaki Island, South Japan", *Japanese Journal of Ecology*, Vol. 33, pp.231-234.

Swaminathan, M.S. and Deshmukh (1995). *Genetic Engineering and Adaptation to Climate Change*. Final Report of the DBT-CRSARD Project, M. S. Swaminathan Research Foundation, Chennai.

Tamai, S., T. Nakasuga, R. Tabuchi and K. Ogino (1986). "Standing Biomass of Mangrove Forests in Southern Thailand", *Journal of the Japanese Forestry Society*, Vol. 68, pp.384-388.

Tang, H.T. *et al.* (1984). "Mangrove Forests of Peninsular Malaysia: A Review of Management and Research Objectives and Priorities", in E. Soepadmo, *et al.* (eds.), *Proceedings of the Asian Symposium on Mangrove Environment and Resource Management*, Kuala Lumpur, University of Malaya.

Tateda, Y. (2002), "Sequestration of CO_2 from Planted Mangroves, New Counter-measure of Global Warming by Restoration of Lost Mangrove Forests", Brief Note, *CRIEPT News*, 367.

Thivakaran, G.A. *et al.* (2002). "Vegetation Structure of Kachchh Mangroves, Gujarat, Northwest Coast of India", *Indian Journal of Marine Sciences*, Vol. 32(1), March 2003, pp.37-44.

Twilley, R.R., R.H. Chen and T. Hargis (1992). "Carbon Sinks in Mangrove and their Implications to Carbon Budget of Tropical Coastal Ecosystems", *Water, Air, and Soil Pollution*, Vol. 64, pp.265-288.

Twilley, R.R., S.C. Snedaker, Yañez-Arancibia and E. Medina (1995). "Mangrove Ecosystems", Ch. 6.1.11, in V.H. Heywood (ed.), *Global Biodiversity Assessment*, UNEP/Cambridge University Press, Cambridge, UK.

United Nations Environment Programme (1995). *Global Biodiversity Assessment*, Cambridge University Press.

Untawale, A.G. (1980). "Present Status of Mangroves along the Waste Coast of India", in *Mangrove Environment, Research and Management*, University of Malaya, Kuala Lumpur, Malaysia.

———. (1984) "Mangroves of India: Present Status and Multiple uses and Practices", *Status Report Submitted to the UNDP/UNESCO Regional Mangrove project for Asia and the Pacific*, p.28.

———. (1984). "Present Status of Mangroves Along the West Coast of India", Asian Symposium on *Mangrove Environment, Research and Management*, Proceedings, pp.57-74.

———. (1985). "Exploitation of Mangroves in India", *Mangrove Ecosystem of Asia and Pacific—Status, Exploitation and Management*, pp.220-226.

———. (1986). *How to Grow Mangroves?*, National Institute of Oceanography, Goa, p.9.

———. (1987). "Some Observations on Mangroves Afforestation along Central West Coast of India", *Proceedings of the International Symposium on Afforestation of Salt Affected Soils*, Vol. 2, pp.133-150.

———. (1987). "Conservation in Indian Mangroves: A National Perspective", in T.S.S. Rao, R. Natarajan, B.N. Desai, G. Narayanaswamy and S.R. Bhat (eds.), *A Special Collection of Papers tp Felicitate Dr. S.Z. Qasim on his 60th Birthday*, pp.85-104.

———. (1991). "Research and Management of Mangrove Ecosystems in India", in S.V. Deshmukh, R. Mahalingam (eds.), *A Global Network of Mangrove Genetic Resource Centres*, Project Formulation Workshop, Centre for Research on Sustainable Agriculture and Rural Development, Madras, pp.27-33.

Untawale, A.G and Sayeeda Wafer (1991). "Socio-economic Significance of Mangroves in India", Proc. Symp. on *Significance of Mangroves*, pp.5-7.

Untawale, A.G, Sayeeda Wafer and M.Wafer (1997). "Ecological Value of Sacred Mangroves of Biodiversity Conservation in India", from *The Role of Sacred Groves in Conservation and Management of Biological Diversity.* pp.292-306.

Untawale, A.G. and A.H. Parulekar (1978). *Status Report on Pirotan Island (Gulf of Kuchchh) for Starting a Conservation Plan,* National Institute of Oceanography, Goa, Report, pp.1-28.

Untawale, A.G. and T.G. Jagtap (1998). "Mangroves, No Wasteland", *The Hindu Survey of Indian Environment,* Vol. 98, pp.71-74.

WRI (1996). *World Resources 1996-99: A Guide to the Global Environment.* WRI/UNEP/UNDP/ WB, Oxford University Press, New York, pp.342.

WWF (1992). "Coastal Marine Ecosystems and Anthropogenic Pressure in the Gulf of Kachchh", *A Report on the Colloquium,* WWF-Rajkot Division, India.

Web References

http://www.aims.gov.au/pages/publications.html

http://www.biology.eku.edu/SCHUSTER/bio%20802/mangrove.htm

http://www.bsu.edu/eft/belize/imangrove.html

http://www.cep.unep.org/pubs/Techreports/tr04en/chapter4.htm

http://www.earthisland.org/eijournal/journal.html

http://www.fao.org

http://fisheries.gujarat.gov.in/fishataglance.htm

http://forest.and.nic.in/frst-mangroves1.htm

http://www.indian-ocean.org/bioinformatics/mangrove.html

http://landbase.hq.unu.edu/Conference/Abstracts/Kathiresan.htm

http://www.lee-county.com/DCD/Environmental/Mangroves.htm

http://www.mangroveindia.org

http://www.mangroves.org

http://www.mangrovesdirect.com

http://mangrove.nus.edu.sg/guidebooks/text/1002.htm

http://www.nationalgeographic.com/ngm/0702/index.html

http://www.oneocean.org/index.html

http://www.oceanoasis.org/fieldguide/avic-ger.html

http://www.pcedindia.com/peoplescomm/coastalecosystem_2a.htm

http://www.pemsea.org/young%20environ/ye101/mangrove1.htm

http://www.ramsar.org/values_intro_e.htm

http://www.wrm.org.uy/bulletin/51/mangroves.html

A